FOSSILS

INSIDE OUT

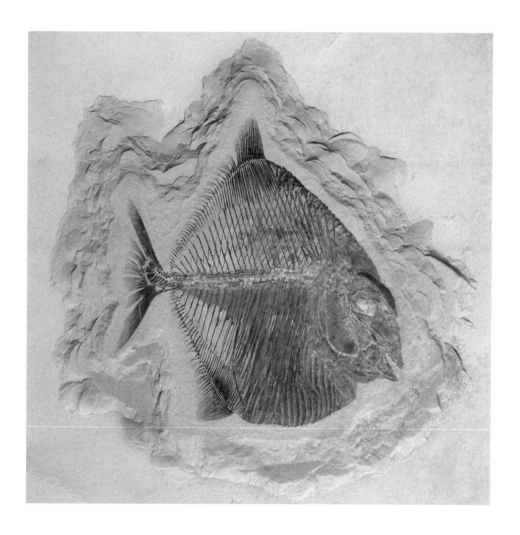

Wild Horizons® Publishing, Inc.

Tucson, Arizona

MOONFISH

Previous page: An exquisite fossil of *Paleobalistum goedeli*, a bony fish that lived in shallow tropical seas when large dinosaurs ruled the land late in the Mesozoic Era. Moonfish specimens have been found in Lebanon; this one is 7.5 inches (19 cm) long—courtesy of Stefano Piccini/Geoworld Group.

PETRIFIED FOREST

Logs and fragments of agatized wood continue to emerge from the heavily eroded desert landscape in northern Arizona's Petrified Forest National Park. About 220 million years ago (mya), this was a vast river basin with galleries of giant coniferous trees, including *Araucarioxylon arizonicum* (**seen here**). Humans who populated the area after the last Ice Age—around 13,000 years ago—made stone tools from petrified wood.

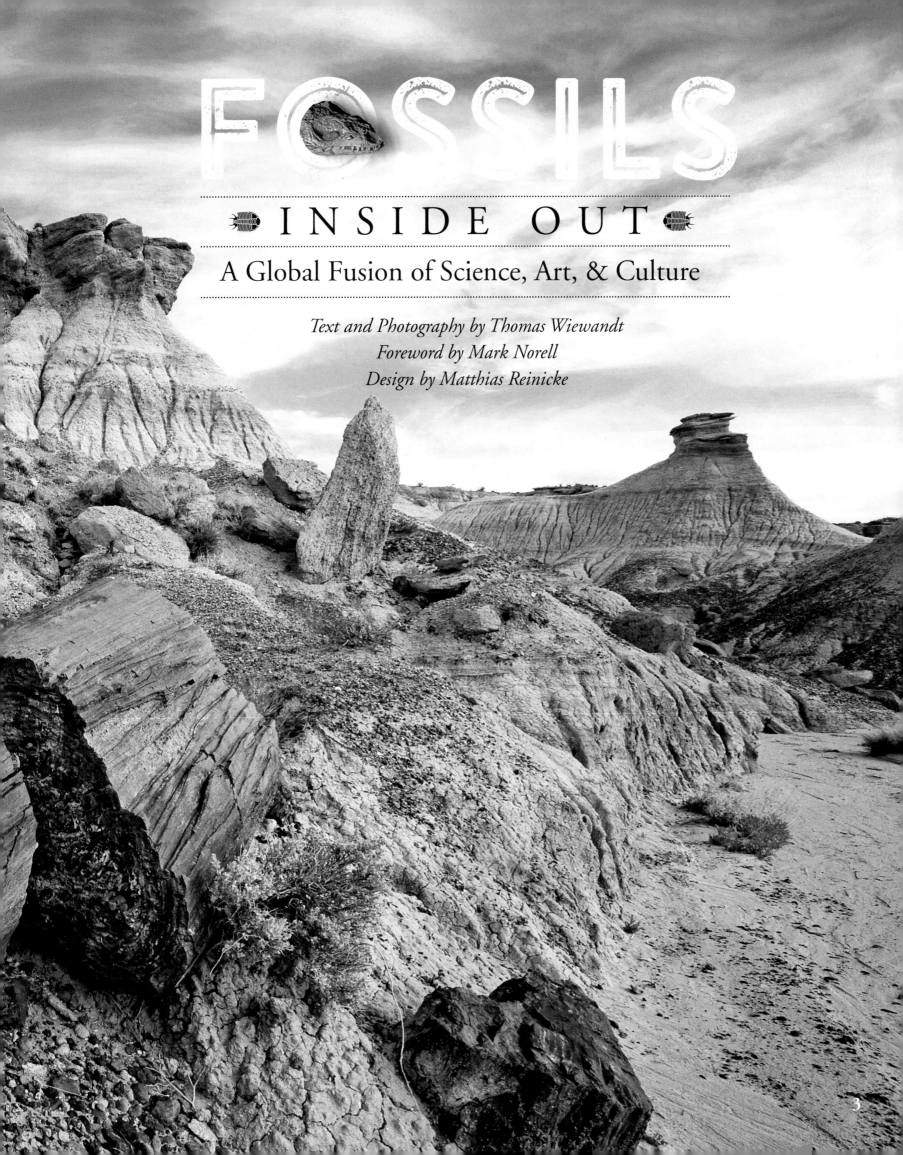

FOSSILS

⊱ INSIDE OUT ⊰

A Global Fusion of Science, Art, & Culture

Text and Photography by Thomas Wiewandt
Foreword by Mark Norell
Design by Matthias Reinicke

Ernst Haeckel

© 2021 Wild Horizons Publishing Inc.
First Edition 2021   10 9 8 7 6 5 4 3 2 1
Printed in Canada
ISBN-13: 978-1-879728-08-0
Publisher's Cataloging-in-Publication Data:

Names: Wiewandt, Thomas A. (Thomas Alan), author, photographer. |
    Norell, Mark, writer of foreword.
Title: Fossils inside out : a global fusion of science, art, & culture / text and
    photography by Thomas Wiewandt ; foreword by Mark Norell. ;
    design by Matthias Reinicke.
Description: Tucson : Wild Horizons, 2021. | Includes bibliographical
    references and index.
Identifiers: LCCN 2020919008 (print) | ISBN 978-1-879728-08-0
    (hardcover)
Subjects: LCSH: Fossils. | Paleontology. | Prehistoric animals. |
    Natural history. | Evolution. | Illustrated books. |
    BISAC: NATURE / Fossils. | SCIENCE / Paleontology. |
    SCIENCE / Natural History. | SCIENCE / Life Sciences / Evolution.
Classification: LCC QE714.5 .W54 2021 (print) | LCC QE714.5 (ebook) |
    DDC 560--dc23.

Includes list of recommended books and index.

Wild Horizons® Publishing, dba Wild Horizons Productions
5757 W. Sweetwater Drive
Tucson, Arizona 85745-9083 USA
Tel: 520-743-4551; Fax. 520-743-4552
Website Contact: http://www.wildhorizons.com

This book is available at discounted prices for bulk purchases in the
United States by corporations, institutions, and other organizations.

TREE OF LIFE

When ships set sail to explore the world—before the airplane was invented in 1903—many seafaring expeditions included scientists. They documented the natural history of new-found places with meticulously kept journals and specimen collections. British biologist-geologist Charles Darwin, British biologist-geographer-social critic Alfred Russel Wallace, and German biologist-physician-artist Ernst Haeckel were among them. They believed that the world could be best understood through scientific inquiry. They were keen observers, interdisciplinary thinkers, and true naturalists who spent decades amassing data. All three men did their best to identify relationships between species and groups of species as they changed through time—*phylogeny*—a term coined by Haeckel in 1866. He was also the first to use the word *ecology*.

Ernst Haeckel was trained as a physician but stopped practicing medicine in 1859 after reading Darwin's *Origin of Species*. He was an advocate of Darwin's (and Wallace's) evolutionary theory and began constructing phylogenetic "trees" as a way to visualize relationships among groups of organisms—he made trees for nearly everything he studied. Darwin never represented groups of plants and animals visually the way Haeckel did. And in those days, "Man" was considered the pinnacle achievement of evolution, and Haeckel's trees reflected this notion.

Haeckel's 1910 *Paleontological Tree of Vertebrates* (**right**) is actually quite remarkable. He aligned this bushy tree of vertebrate life with a geological time scale based on fossil records from geologists. Although now much refined in the shadow of new discoveries, Haeckel's trees of life sketched more than 100 years ago have had a lasting influence on visual representations of evolutionary relationships. As you read the chapters to come—especially the sections on lobe-finned fish (in Chapter 2) and the evolution of birds (in Chapter 6)—keep this diagram in mind.

Although similar in concept, phylogenetic "trees" are now being replaced with *clade diagrams* (= *cladograms*). Simply put, a clade is a natural grouping of organisms with a common ancestor, similar to one branch of a phylogenetic tree with its descendants on branchlets—typically presented as a connected series of straight lines, more like stick-figures.

Swedish biologist-physician Carolus Linnaeus (1707-1778) developed a classification system for naming and ranking organisms. Names in those days were long and unwieldy—the tomato, for example, had been given a seven-word descriptive name! To simplify and standardize names, he came up with a *binomial* system with two-part names, still in use today—*Homo sapiens*, for example. And he organized species with shared anatomical features into species groups—*genera* (plural of genus). So the first part of the name is the species group (genus) and the second part the organism's species. He grouped similar genera into *orders*, similar orders into *classes*, and similar classes into *kingdoms*—a heirarchy of relationships. Today, this ranking system has been expanded to include *families* between genera and orders, and *phyla* (plural of phylum) between classes and kingdoms, and some newer, higher-level groupings. Each level is called a *taxon*.

Linnaeus recognized two kingdoms of life—Plantae and Animalia—and in 1866, Haeckel added a third, *Protista*, which he applied to all single-celled organisms and "lower" fungi, like molds. Throughout the early history of its use, Kingdom Protista has been a "wastebasket" kingdom—any life form that didn't comfortably fit with plants or animals was tossed into the Protista. Fast forward to 1969, Robert H. Whittaker at Cornell University, NY, proposed five kingdoms, still widely used today, especially for teaching basic biology: Animalia, Plantae, Fungi, Protista, and Monera (for bacteria; a taxon first used as a phylum by Haeckel in 1866). But as scientists have become better equipped with new technologies to examine single-celled organisms and DNA, classification systems continue to be subdivided, refined, and reorganized.

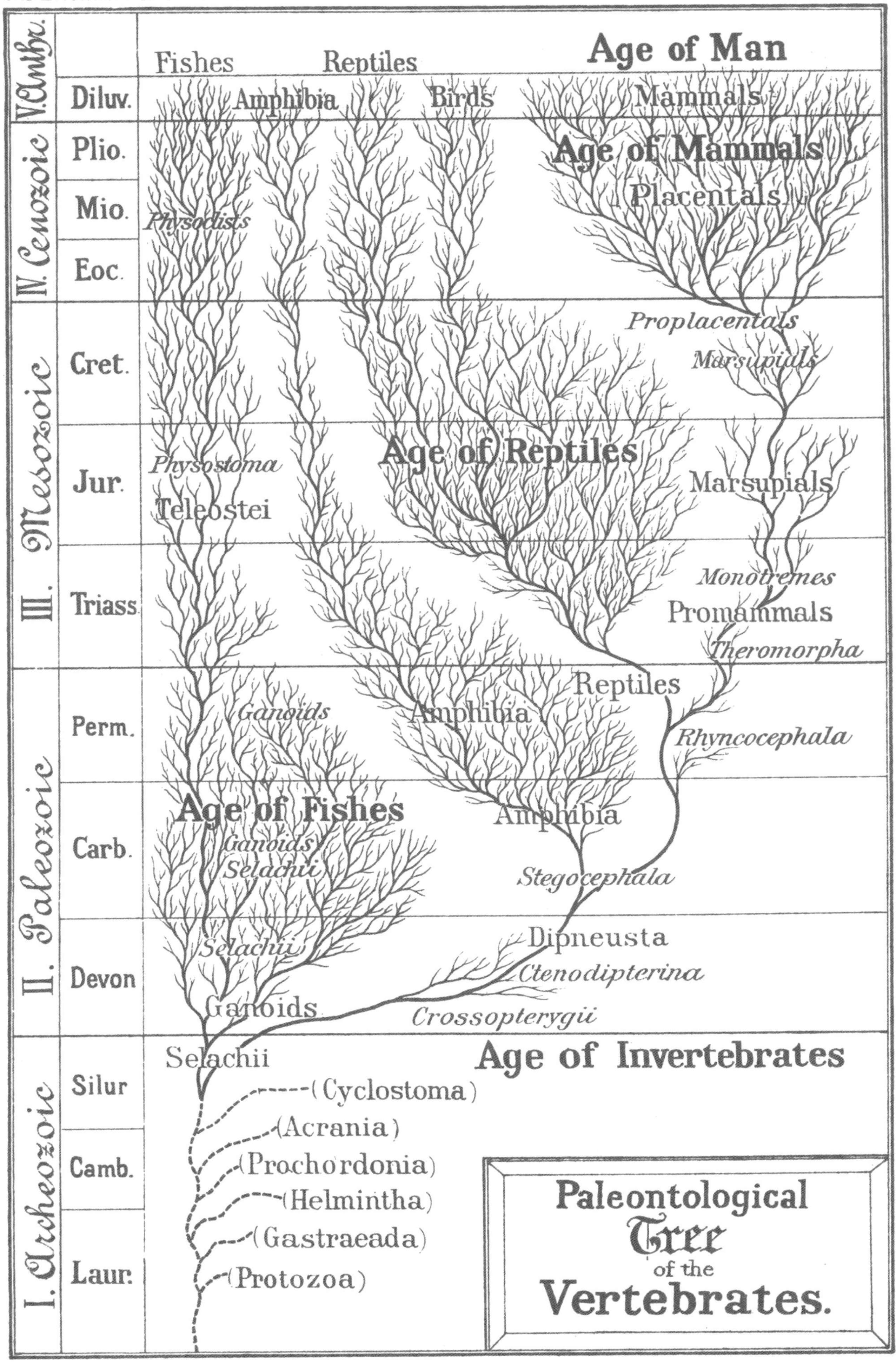

V. Anthr.		
	Diluv.	
IV. Cenozoic	Plio.	
	Mio.	
	Eoc.	
III. Mesozoic	Cret.	
	Jur.	
	Triass.	
II. Paleozoic	Perm.	
	Carb.	
	Devon.	
I. Archeozoic	Silur.	
	Camb.	
	Laur.	

Fishes   Reptiles

Age of Man

Amphibia   Birds   Mammals

Age of Mammals

Placentals

Physoclists

Proplacentals

Marsupials

Physostoma

Age of Reptiles

Marsupials

Teleostei

Monotremes

Promammals

Theromorpha

Reptiles

Ganoids   Amphibia

Rhyncocephala

Age of Fishes

Amphibia

Ganoids
Selachii

Stegocephala

Selachii   Dipneusta

Ctenodipterina

Ganoids   *Crossopterygii*

Selachii   **Age of Invertebrates**

(Cyclostoma)

(Acrania)

(Prochordonia)

(Helmintha)

(Gastraeada)

(Protozoa)

Paleontological Tree of the Vertebrates.

E. Haeckel del.

5

CONTENTS

1 DEALERS TO DINOSAURS: EVOLUTION OF WORLD'S GREATEST FOSSIL SHOW   PAGE 12

Historical perspectives on Tucson's annual Gem, Mineral, & Fossil Showcase • Tucson & Gem & Mineral Society • Tucson Convention Center show • Association of Applied Paleontological Sciences • Tucson's satellite mineral & fossils shows • fossil show trends from 1987-2020

2 FOSSILS : MESSENGERS FROM THE PAST   PAGE 24

Kinds of fossils • uncertain journey from deposition to preservation • pseudofossils • examples of "living fossils," a dated concept—lobe-finned fish, ginkgo trees, & crocodilians

3 FOSSILS LOST & FOUND   PAGE 36

Environments that favor and don't favor fossilization • where fossils are found—desert badlands; frozen places & dry caves/mummification; sinkholes & cenotes; tar pits; tree resin/amber

4 GEOLOGICAL TIME & DRIFTING CONTINENTS   PAGE 46

Geological time scale/history of life chart • eight techniques for dating fossils/relative & absolute • historical perspectives on continental drift • eight time slices of continents in motion from 1.4 billion years ago to present • fossils from Gondwana supercontinent • map of today's tectonic plates

5 COLLECTING FOSSILS   PAGE 58

Who collects fossils, what can be collected, and where • hunting fossils in remote places • fossil quarries • international fossil trade and problematic policies • lessons from catastrophic museum fires • legal guidelines • field experiences for amateur collectors

SEED FERNS

Background image: Vast swamp forests were widespread during the Carboniferous Period (359–299 mya), when much of the world's climate was warm and humid. Tropical plants that existed in North America, Europe, and Asia at the time became today's coal deposits. Giant seed ferns were among the dominant trees. In this tapestry of fossilized seed ferns found near St. Clair, Pennsylvania, USA, the leaves were replaced with a white clay mineral during fossilization—part of an enormous slab on display in the Prehistoric Journey exhibit at the Denver Museum of Nature and Science in Colorado.

There are all kinds of collectors and all kinds of collections. My day job is chairing the Division of Paleontology at the American Museum of Natural History in New York City, the home of one of the largest fossil collections in the world. Our specimens range from shells of single-celled organisms to skeletons of large dinosaurs. Some are iconic, like the type specimen of *Tyrannosaurus rex*. This is an academic collection visited by hundreds of scientists each year. They measure, scan, photograph, compare and observe. It is a living collection used to test hypotheses to understand the history of life on our planet. It is also a display collection in a museum accustomed to welcoming more than 4 million visitors a year.

We know from visitor surveys that the number one attraction at the AMNH is the fossils—specifically, the dinosaurs. These fossils have inspired future scientists, visual artists, writers, daydreamers, and other creative souls. They fascinate both adults and children. Such visitors, who hail from all over the globe, do not come to study the anatomical minutiae that occupy the minds of professional paleontologists. Instead there is something captivating here. Perhaps it has to do with the immensity of time—when and how long ago did these creatures live? Other reasons are aesthetic. Anyone looking at the iridescent hues of a fossil ammonite, the wild spines of a trilobite, or the 11-inch tooth of a *Tyrannosaurus rex* has to be amazed.

Fossils have been collected for a long time. We know that the ancient Egyptians, Greeks, and Chinese observed and collected fossils. In the modern era, before the time of natural history museums, collecting fossils was a gentlemanly sport, especially in Western Europe. Many fossils were incorporated into "cabinets of curiosities"; others became the triggers for local myths, and some were housed in local churches. The famous Klagenfurt dragon myth was supported by bones of a fossilized creature found in an Austrian quarry in 1335. But the dragon's head turned out to be the skull of an extinct woolly rhinoceros. Himalayan ammonites are considered symbols of Vishnu or bolts of lightning.

Even thousands of years ago, fossils were important scientifically. People began to notice things that now seem obvious: fossil shells found on mountain tops indicated that the land was once under water. The presence of fossil bamboo in cold areas showed the climate had been different in the past. And more recent observations of similar fossils found in Antarctica, South America, and Africa suggested that these continents were once joined into a super-continent.

I first attended the Tucson Gem and Mineral Show when I was in high school in the early 1970s. It was rocks and more rocks, some with fossils—the ubiquitous Green River fish, agatized Morrison dinosaur bones destined to be cut and polished into bolo ties and belt buckles, and some trilobites mostly from Nevada and Utah. Much of this inventory was laid out in swap-meet style, in vacant lots. About the time I finished graduate school in the late 1980s, the show exploded. Suddenly there was a lot of interest and a lot of money was involved, the foundation of a collector's market.

The lower end was driven by the curious and decorators—think shark teeth, fossil fish, and mollusks. But as in art, there is a higher end. And it became not uncommon to see specimens on the market for six figures or even seven. Some of these were purchased by museums, but many went into private hands. What had once been the activity of nerds, weirdos, desert rats, and scout troop leaders was now the focus of hedge fund managers, actors, lawyers, royalty, and heirs.

This competition from the private sector drove some academic and museum paleontologists apoplectic. They argued that monetization of fossils encouraged looting from public lands and would result in losses to paleontology. As a curator, I cannot with a straight face deny that unethical dealers and collectors exist. However, this is not the norm. People want to own fossils for a range of reasons, not all of them scientific. But in my experience both collectors and dealers are very interested in the pursuit of knowledge from professionals.

Many people in the non-scientific arena have been overly generous in donating to science and giving professionals access to important specimens. I hope we see more of this, as long as specimens are collected legally and ethically, international law is observed, and the specimens are well cared for. As the market and private collections grow, collectors and academics can develop mutually beneficial relationships, like in the art world.

That is what makes this book unique. Featuring several aspects of paleontological science, much gathered from non-academic collectors, it brilliantly covers the evolution of life and the planet. It also provides excellent images of fantastic never-before-seen fossils, accompanied by a lively narrative. It's a beautiful book that will certainly sit on the table in my office!

Mark Norell, PhD
Macaulay Curator of Paleontology
Chairman, Division of Paleontology
American Museum of Natural History

As a child I never knew what my father did for a living—not exactly anyway. Those were strange times, and none of us kids knew what our fathers were doing. Mine was an engineer, good enough. Such was life for children growing up in Los Alamos, New Mexico, birthplace of the atomic bomb, developed between 1943 and 1945 in the town's top-secret nuclear weapons lab. Our family moved to Los Alamos—commonly known as "the Hill"—in 1948, when the population was about 7,500. Only after my father's death did I learned that because of his natural talent for mathematics, he had been asked to recalculate equations behind the first atomic bomb. Wow! Even in post-war years, nuclear scientists couldn't fully comprehend what they had unleashed.

Until 1957, "Lost Almost" was a federally gated community, and everyone lived in government-owned housing, a compound for physicists and engineers, where kids were encouraged to explore new ideas and to break bounds. Individualism ruled supreme. Schools were excellent, and science fairs were big. No one locked their doors. And the town's idyllic mountain setting nurtured my innate fascination with wild places. Kids wandered freely in pine-studded mountains and rugged canyons. We had family cookouts in Bandelier National Monument and could explore Indian markets and rock shops in Santa Fe.

Where any of us live at an early age comes down to "luck of the draw," and I'll be forever grateful to have spent my childhood at the right place and time for a head start in life. More likely than not, this is where I gained the desire and courage to pursue an interdisciplinary career. A deeper understanding of living things helped fuel my creative passions, and the medium of photography allowed me to blend science and art as a storyteller. I'm inspired by the mystery, beauty, and complexity of the natural world, in both form and function.

While an undergraduate at Marietta College in Ohio, a liberal arts school, I studied everything from comparative anatomy and English literature to studio art and music appreciation. During those years—after two summers at a field research station owned by the American Museum of Natural History—I abandoned the idea of applying to medical school. Instead, I entered a graduate program at the University of Arizona to study desert ecology, followed by a move to Cornell three years later to further my education in their Department of Ecology and Evolutionary Biology. My research interests took me to the remote Caribbean island of Mona where I lived for three years, immersed in field studies of iguanas, hermit crabs, sea turtles, introduced mammals, and more. University professors, close friends, and colleagues in the Puerto Rican Department of Natural Resources opened their doors and hearts to this wayward naturalist. My goals were not defined by financial security, and twice, although armed with a PhD, I ended up on food stamps—a humbling experience. After documenting the natural history of Mona's native land

iguana for my doctoral research, I completed a 16mm film about it with assistance from the BBC—scripted and narrated by Sir David Attenborough. This led to more opportunities in filmmaking, including three educational films for the National Geographic Society. Technology was changing fast at that time—first the move to Super-16 and then to video, accompanied by a lot of new production equipment, more than I could afford or even wanted. I struggle with new gadgetry and shun working with large crews under tight production deadlines, so I decided to steer my career in another direction.

I began a photographic safari and workshop business called Wild Horizons®, founded to enhance environmental appreciation through nature photography, designed to give others an opportunity to step behind the scenes—not only to see, but to hear, smell, and feel wild places.

My travels have taken me to many wonderful locales, but by 1983 I had decided to return permanently to the Sonoran Desert. And in 1986 I began exploring Tucson's annual Gem and Mineral Show and its satellite fossil shows, described in the following chapter. I was stunned by the beauty of geological and paleontological treasures flown and trucked in from every corner of the globe. Fossils, in particular, caught my eye. So I set up shop on a card table in a pub at the motel where most fossil specimens were on display, adjacent to the showroom of the Black Hills Institute of Geological Research. Thanks to the generosity of the Black Hills team and contacts made with other fossil dealers, friendships grew that led to a feature article in *Audubon* magazine. This in turn fueled my interest in publishing—the beginning of Wild Horizons Publishing, aka Wild Horizons Productions.

Many of my tours and workshops focused on the American West, a region filled with captivating landscapes. My clients browsed through books available in gift shops, but most were specific to individual parks or dense with geological jargon. So I saw the need for a synthesis that would appeal to the average traveler. This led to my first big publishing project in 2001—*The Southwest Inside Out: An Illustrated Guide to the Land and Its History*, now in its fourth edition.

Then in 2005, after purchasing my first digital camera, I returned to independent filmmaking, and while attending a fundraiser for a new desert park, I met guest musician Gary Stroutsos. He suggested we collaborate on an audio-visual project—the birth of *Desert Dreams: Celebrating Five Seasons in the Sonoran Desert* in 2012—based on my 1990 children's book *Hidden Life of the Desert* (a greatly expanded edition was published by Mountain Press in 2010). With help from KAET-TV in Phoenix, this documentary has aired on national public television stations as a fundraising program for six consecutive years.

Now comes my third big labor of love, *Fossils Inside Out*. This project is rooted in 34 years of photography, field work, and relationships established with fossil collectors, paleontologists, paleo-artists, and specimen preparators I've met in Tucson at what has become the Gem, Mineral, & Fossil Showcase. Since the mid-1980s I have been photographing the best specimens to appear at the show before they are sold. Without these folks, there would be no book, and I'm deeply indebted to them for their trust, camaraderie, and assistance along the way. Many individuals, companies, institutions, and government agencies have granted me permission to publish images of their specimens, paintings, artwork, or photographs, as credited in my captions—thank you all!

Beyond the Tucson show, a number of individuals, museums, parks, and other organizations deserve a special call-out for offering me privileged access to their specimens, displays, and behind-the-scenes activities, often by inviting me into their homes, fossil quarries, paleo prep-labs, and studios: Arvid Aase, Fossil Butte National Monument Visitor Center Museum; Bill Barker, Sahara Sea Collection; Denver Museum of Nature & Science staff; Dinosaur National Monument Quarry Exhibit staff; the Hebdon clan, Warfield Fossils; Rick Hunter, Museum of Ancient Life/Thanksgiving Point Institute; Tom Kaye, Foundation for Scientific Advancement; Peter Larson & colleagues, Black Hills Institute of Geological Research & Museum; Tom Lindgren, GeoDécor; Jerome Montgomery, Green River Stone Company; Stefano Piccini, GeoWorld; Burkhard Pohl, Wyoming Dinosaur Center; Brock Sisson, Fossilogic; Mike Triebold, Rocky Mountain Dinosaur Resource Center; Justin Wilkins, Mammoth Site & Museum of Hot Springs, SD; and Scott Williams of Petrified Forest National Park (in 2006, when I was a park Artist in Residence). In 2002, when I visited the American Museum of Natural History to photograph a fossilized skull for Dan Beck's 2005 treatise on the *Biology of Gila Monsters and Beaded Lizards*, I connected with paleontologist Mark Norell and am especially grateful that he agreed to write the Foreword to this book.

In 1995 I met Barry B. Brown, an aspiring photographer based in South Dakota who was collecting and preparing fossil specimens at the time. His talent was obvious, so I encouraged him to pursue a career in photography. But then he vanished—and friends later told me that his wife Aimee was now training dolphins in Curacao, and Barry had begun underwater photography. We reconnected, the beginning of an enduring friendship and his growth as a professional photographer. Barry is as comfortable in the ocean as a fish and has become a master with digital cameras and lighting equipment. I am honored to feature several of his stunning images in the pages ahead.

Paleontologist Serge Xerri, another talented photographer, brings magnificent vertebrate skeletons from Morocco to Tucson. He, too, has generously donated several outstanding photos to this project—thank you, Serge.

The 292 images without photo credits (90% of the total) reside in my personal archive. As a member of the Iguana Specialist Group of the International Union for the Conservation of Nature (IUCN), I have had many travel opportunities before and after our ISG annual conferences in the Caribbean region since 1993, all documented photographically. Some of these images—mostly landscapes—have been useful here, in unanticipated ways. I'm grateful to our hosting countries and colleagues for their warm hospitality and guidance to places of special interest.

Because of the broad geographic and editorial scope of this book and my intent to make it accurate but non-technical, many creative and scholarly individuals were vital to its development. Sally Antrobus, a professional copy-editor, has become a close friend and an essential part of my workflow over the years; she edits with care and sensitivity, found only in the best editors. And Jon Tennant, a gifted young British paleontologist (on opposite page), scrutinized the text for scientific relevance and accuracy. Working with him was a joy, and his comments were extremely helpful. Thank you Sally and Jon.

Two trusted readers with different backgrounds—outdoor educator Luann Sewell Waters and filmmaker Jeffrey Cravath—reviewed the entire text for clarity, while Nan Rollings, Carol Townsend, and Brock Sisson offered constructive comments on individual chapters. Several scientists also helped as content consultants: paleontologists Henry Galiano, Frederic Lacombat, Bill Barker, and Max C. Langer; plus imaging specialist Tom Kaye; herpetologist Charles J. Cole; geologist James St. John; and mineralogist Debbie Colodner. I am extremely grateful to all for making this a better book.

Writing about the *Art of Fossil Preparation & Display* (Chapter 9) required inside perspectives not to be found in print or online. More than anyone, Brock Sisson, owner of Fossilogic, guided me through his profession and outlined key points for my text—a huge thank you and well-deserved accolades for skills and generosity go to Brock as a friend and advisor. Terry Chase, owner of the world-class paleo-exhibit design company Chase Studio, also shared his experience and insights. Thank you Terry, and best wishes for success with your book on this subject. Other contributors are credited in the text.

Great design is crucial for a publication to be inviting, lively, and easy to read. I have been extremely fortunate to have connected with two of the best designers in the business, Carol Haralson—designer of *The Southwest Inside Out*—and Matthias Reinicke, who designed this volume. Carol, now retired, came up with the "*Inside Out*" brand for my Southwest book, wording I have retained for *Fossils Inside Out*. Both Matthias and Carol were a pleasure to work with, in no small part because they were willing to let me participate in the design process throughout. I have been adamant about keeping every page spread as self-contained as possible for "stop-and-go" reading. I'm deeply indebted to both designers for their patience, friendship, and professionalism. They have done more than words can express to bring these visions to life.

Matthias is also a gifted graphic designer with much experience in creating interpretive artwork for museum exhibits. He and three other artists—Paul Mirocha, Mark Witton, and David Fischer—have contributed engaging, easy-to-understand illustrations to this book. And the up-to-date maps showing continental movements through deep time were customized for us by paleogeographer Andrew Merdith, of University Claude Bernard Lyon 1, France. Thank you all for your skillful translations of complex scientific information into beautiful, easy-to-grasp illustrations.

Producing almost anything nowadays requires computer tech support—especially for projects masterminded by those of us who don't have what it takes to conquer the digital bugs ("trilobytes") and challenges that plague us all. So I'll always be thankful for the wizardry of Michael Robinson, who keeps my boat afloat, and Sandy Binns who helps to maintain my websites. I am also grateful to Marilyn Anderson and Jeffrey Cravath for their assistance with the index.

Many other colleagues and friends have contributed to this project in significant ways, and I genuinely appreciate all the support and good faith sent my way during this journey.

IN MEMORY OF

photo: EGU Media Library

JONATHAN P. TENNANT
(1988 - 2020)

To shine, nearly all writers need good editors—copy-editors and in some cases content editors. While writing this book, I began hunting for a content editor, specifically a talented paleontologist with experience in communicating with the general public. I quickly connected with self-described "Rogue Paleontologist" Jonathan Tennant, who received his PhD in 2017 at Imperial College London, where he was awarded the Janet Watson Prize for research excellence and good citizenship. His research focused on diversity and extinction patterns of land animals during the Age of Reptiles, the Mesozoic Era. Even before graduating, Jon had authored a children's book about dinosaurs, and had another one on the way in 2020. He wrote popular articles for magazines too—*BBC Wildlife, Discover, Earth Touch News Network,* and *The Conversation*.

For such a young scientist, Jon had an exceptional record of research publications, all of which he made freely available through *Open Access,* an important movement to make scientific research and data available to everyone—professionals and citizens alike—via modern digital technologies for educational outreach. Achieving this requires cultural change, a shift from traditional publishing in scientific journals to sharing information publicly, often while research is in progress. This concept is now embraced by the European Geosciences Union (EGU) and many other international organizations.

Jon was a passionate spokesperson for what has become generally known as *Open Scholarship,* to promote interdisciplinary cooperation and better transparency in science, while challenging companies who lock-up publicly funded research behind paywalls for private gain. He founded a digital publishing platform—*PaleorXiv*—and a peer-to-peer community for open research practices—*Open Science MOOC*. Jon was also an ambassador for the Center for Open Science, and Editor for the journals *Geoscience Communication* and *Evidence-Based Healthcare*. In 2018 he received the Jean-Claude Guedon Prize, rewarding "the best article on the issues of scholarly publications and/or open access." And as an independent researcher affiliated with the Institute for Globally Distributed Open Research and Education (IGDORE), he won a travel award to work with them in Bali.

As a free spirit who prefers to blaze trails "outside the box," I felt a connection with Jon right from the start. He was always down-to-earth and a true professional, so communicating with him for nine months via email was a rare pleasure. Besides, who wouldn't like a guy with a blog named *Green Tea and Velociraptors*! Whether on 2019 speaking engagements in Europe, Peru, or Bali, Jon was always accessible. And his genuine concern for the future of our planet came across in his editing.

I chanced across Jon's obituary on EGU's website six months after his fatal motorcycle accident in Bali on 9 April 2020. The news hit me like a freight train. The world has lost a rising star—a phenomenal scholar, a champion communicator, an amazing human being—struck down in his prime at 32. It's my hope that today's youth will honor Jon's life by digging beneath the surface of what they read and hear, to become astute observers and critical thinkers—the very foundation of science. Jon fervently believed that public understanding of science is crucial to the future of humanity and urged us all to "engage in civil, constructive, and thoughtful conversations." Jon also saw many potential leaders lurking in the shadows of the internet and urged them to step forward, to become engaged in the real world ... to carry the torch!

DEALERS
TO DINOSAURS

EVOLUTION OF WORLD'S GREATEST FOSSIL SHOW

1

Great things spring from humble beginnings. Who would have guessed that a 1955 display of several thousand beautiful and unusual rocks at a local elementary school would evolve into the world's largest showcase for gems, minerals, and fossils.

A reporter from the Arizona Daily Star wrote that 1,300 mineral collectors and rock hounds flocked to this two-day event staged by a local club, the **Tucson Gem & Mineral Society (TGMS)**. But even back then, the show had an international flavor. It attracted attendees from most states in the USA, Canada, and Mexico. Uranium rocks on display came from the American West, Canada, England, Norway, Germany, and what was then Czechoslovakia. And the Geiger counter demonstration was a hit.

COPPER MINERALS

Tapestry of copper-bearing minerals—mostly malachite and azurite—in a boulder from an Arizona mine. The world's leading producer of copper is Chile, followed by the USA and Peru.

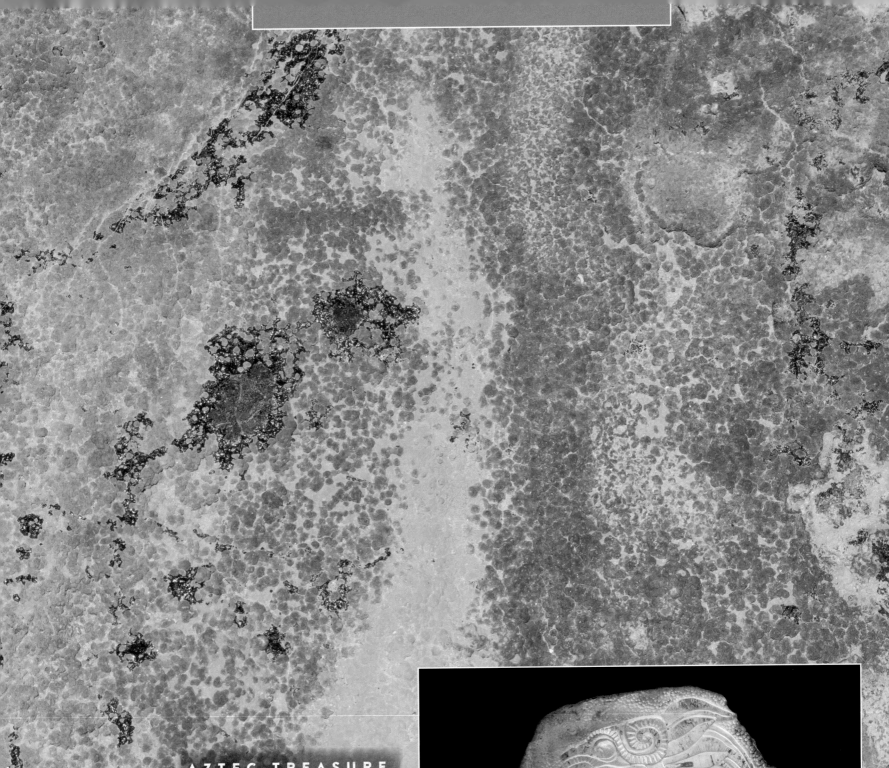

AZTEC TREASURE

Two amber lip ornaments and two brick-sized blocks of clear amber are among the cultural artifacts listed in an ancient book of the Aztec. They came from *Socanusca*—now in the state of Chiapas, Mexico. Today, amber is mined in Chiapas and Coahuila, but Chiapas remains the hub of **Mexican amber** trade.

Right: A large piece of golden amber with a *Quetzalcoatl*—Feathered Serpent deity of the Aztec—carved on its surface. This piece from Chiapas was part of an educational display at the Tucson Gem and Mineral Society's 2017 show at the Tucson Convention Center. Specimen donated to the TGMS by Forrest & Barbara Cureton.

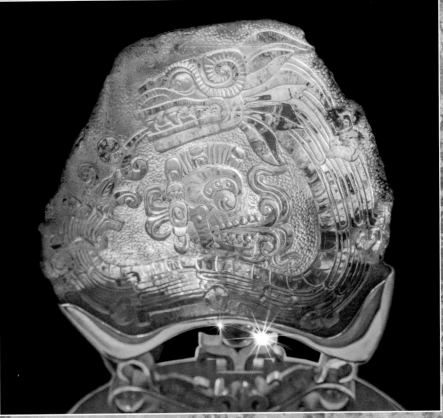

The show became an annual event, and after a successful 15-year-run from a rusty Quonset hut on the Pima County Fair and Rodeo Grounds, it finally moved in 1972 to the newly constructed Tucson Convention Center. The main show, which has continued to focus on precious gems and minerals, has become a four-day event at the Convention Center, typically in mid-February. Each year, the TGMS selects a theme mineral, mineral group, or country. They organize educational seminars and eye-popping displays from private collections and major museums like the Smithsonian Institution. About 250 booths from a wide range of North American and international dealers populate the rest of the exhibit space—to participate, dealers must apply and be invited by the TGMS.

Over the years, "the Tucson show" has undergone a dramatic transformation into a showcase event, with the main show anchoring a series of related ones that cover a wide range of other interests. These "satellite shows" start in the last week of January and end with the four-day TGMS show at the Convention Center in mid-February. Specialized shows have been launched in nearly every major and minor hotel in the city. Vendors sell from their hotel bedrooms, from ad hoc street-side shops, and even from their vehicles. An African Arts Village offers handcrafted items that range from beads and bangles to baskets and blankets. And every January, display tents spring up in empty lots and on county-owned properties, such as the Kino Sports Complex (formerly called Tucson Electric Park). To establish a more permanent presence in Tucson, a few US and international dealers have purchased centrally located homes and warehouse spaces.

A 2019 study documented this trade show's impressive economic impact on Tucson. In 48 independent shows, nearly 5,000 exhibitors arrived from 45 states in the USA and from 42 international locales. Attendance averaged 10,453 customers per show, attracting nearly 65,000 visitors in all. Tucson and its surrounding communities benefited from a total of $200 million in direct expenditures, $81 million of which was for lodging and food. Most shows are open to both wholesalers and the public. Admission is free except to the Convention Center show, and parking is extra in some places. Eighty percent of the dealers feature rocks, minerals, and fossils. People come to buy, sell, and learn about everything from meteorites and dinosaur eggs to gemstones and crystal balls.

HIDDEN COLORS

Above: Fluorescent minerals glow under ultraviolet (= UV) light, colors that are invisible in daylight. These specimens from mines in Sussex County, New Jersey, USA, were on display at the 2008 Tucson Gem and Mineral Show, courtesy of the Sterling Hill Mining Museum. No less than 89 different fluorescent minerals have been found in northwestern New Jersey—some of the finest examples of mineral fluorescence in the world.

SOMETHING FOR EVERYONE

Cut and polished stones in every color, size, and shape imaginable can be purchased at Tucson's city-wide annual Gem, Mineral, & Fossil Showcase event. Minerals abound, but so do fossils, arts, crafts, tool of the trade, books, and more.

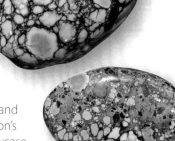

14

TUCSON CONVENTION CENTER SHOW

Two of 14 rows of booths and special exhibits in the main Exhibit Hall of the 2020 show hosted by the Tucson Gem and Mineral Society. Other dealers were assigned space in the Grand Ballroom, Lobby, and Galleria. And meeting rooms booked for guest speakers helped to make this an educational experience for attendees. This was TGMS's 66th show, with the 2020 theme "World Class Minerals."

DEALERS TO DINOSAURS

In 1977 the Tucson showcase gave birth to an important new organization designed to serve as a united voice for the fossil industry: the **American Association of Paleontological Suppliers** (**AAPS**). It was founded in 1978 by a core group of commercial fossil and mineral collectors and preparators. Their mission was to promote ethical collecting practices and to improve lines of communication with the academic and museum paleontological communities. In 1979 the AAPS organized their first fossil show at what was then Tucson's Sheraton Pueblo Inn. Another organization, the **International Association of Paleontological Suppliers** (**IAPS**), founded by a Swiss paleontologist, merged with the AAPS in 2000. Their goal had been to help foreign businesses understand legal matters regarding the import and export of fossils from different countries around the world. Soon thereafter, the AAPS and IAPS incorporated jointly under a new name, the **Association of Applied Paleontological Sciences** (with the same acronym, **AAPS**; see www.AAPS.net).

In 1988, under the management of U.S. Gem Expos, AAPS's mineral and fossil dealers came together in a more cohesive way at the Pueblo Inn. The group adopted the name **Arizona Mineral & Fossil Show** in 1990, and by 1999 participation had reached more than 370 dealers, 135 of them from foreign countries. For the next two decades, hosting hotels changed ownership/names multiple times. Fossil dealers centralized much of their activity at the Hotel Tucson City Center (previously the InnSuites Hotel), and at the **Fossil and Mineral Alley** at Days Inn/Tucson City Center (formerly the Hotel Ramada Ltd.). While at the Ramada, vendors were part of the Arizona Mineral and Fossil Show; but when the hotel switched ownership, the show became an independent entity at the Days Inn.

TENTED SHOW VENUES

Below: Every winter in mid-January—with the approach of Tucson's Gem, Mineral, and Fossil Showcase event—tents spring up in empty lots along city streets. This enormous string of tents by Interstate 10 houses the 22nd Street Show; 2020.

BIG BITE

Administrative Director of AAPS, George Winters, in the gape of a "Fierce Lizard," *Gorgosaurus sp.*, a North American tyrannosaurid dinosaur from Montana, USA. On display in a hotel ballroom at the 2011 Arizona Mineral and Fossil Show, photographed at closing. This skeleton was replicated by the Black Hills Institute of Geological Research.

FOSSILS INSIDE OUT

With the rising need for more display space at the Tucson showcase since the 1980s and '90s, some fossil dealers have purchased buildings, parking lots, and warehouse space for a permanent presence in Tucson. Among them are the **Mineral & Fossil Co-op**, the **Mineral & Fossil Marketplace**, and **Granada Gallery**. A tented venue opened by EonsExpos in 2012—the **22nd Street Mineral, Fossil, Gem, & Jewelry Show**—has grown into a quarter-mile-long string of tents along Interstate 10, with food courts and expanded parking facilities. One tent offers spacious climate-controlled booths with modular art walls, carpeting, and stage lighting; and plans for further expansion continue. All of these shows are within a 2-mile radius of downtown Tucson.

Many dealers miss the camaraderie associated with seeing their colleagues in one central show location. But as a trade-off, Tucson has become the most important place in the world to display, buy, and sell paleontological specimens and art to a global market of museums, academic institutions, shop owners, and private collectors.

DEALERS TO DINOSAURS

MINERAL & FOSSIL CO-OP

All showrooms on this page spread are in the Mineral & Fossil Co-op, an investment by several US and international dealers. These showrooms are open to the public during Tucson's annual event and by appointment at other times.

Two GeoWorld showrooms, photographed in 2020 with permission from owner Stefano Piccini.

Above: Kira Florence admires a woolly mammoth tusk among ammonite and fish fossils, with a cave bear skull in the foreground.

Left: Lizette Perez ponders an assortment of mollusk fossils from France and Italy, and fish from Wyoming's Green River Formation. A tall amethyst geode from Brazil sits in the corner.

Right: Ammonite, *Placenticeras costatum*, found with clam shells, *Inoceramus sagensis* (right)—of Late Cretaceous age (94–66 mya), from South Dakota. Specimen courtesy of Co-op member Neal Larson/Larson Paleontology Unlimited/2019.

Right: Petrified wood furniture and decor in Mineral & Fossil Co-op showroom of Ralph Thompson/Russell-Zuhl Petrified Wood/2015. This cut-and-polished fossilized wood comes from abundant deposits in the American West.

Above: A primitive spiny-finned fish, Order Beryciformes, related to modern squirrelfish and orange roughy; from Late Cretaceous deposits in Lebanon. Specimen courtesy of Stefano Piccini/Geoworld Group/2019.

GeoDecor showroom in the Mineral & Fossil Co-op, photographed in 2016 with permission from owner Tom Lindgren. Specimens on the wall are freshwater turtles and a crocodilian in exceptionally large rock slabs; of Eocene age from the Green River Formation in Wyoming.

DEALERS TO DINOSAURS

As a resident of Tucson, I have had the privilege to observe and photograph the evolution of fossil displays in this trade show since 1988. Apart from its growth, I've noticed some interesting trends and popular fossils:

International Trade: The appearance of new and exciting fossils at the Tucson showcase has changed in character over the years. Rules governing fossil collecting and export are often vague and frustrating to dealers. So what is thought to be legitimate trade one year may be banned the next. Three glaring examples of this have been the importation of fossils from rich deposits in Argentina, Brazil, and China (see Chapter 5).

Replicas: A notable revolution of replicas began in the 1990s. Although replicas of fossils were not new, much of that early effort had been devoted to fabricating missing bones needed to make complete skeletons for museum displays. But now, convincing replicas of one-of-a-kind fossils, including full skeletons of dinosaurs, can be purchased at discounted prices. This trend opened a whole new market for both sellers and buyers that has continued to grow (see Chapter 9).

Amber: For many years amber from Russia, the Dominican Republic, and Mexico has been widespread across the spectrum of Tucson's fossil and jewelry shows. But buyers beware: fake amber has also been on the rise (see Chapter 9).

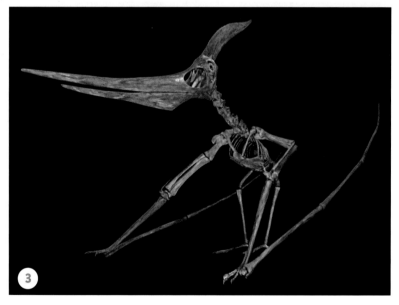

(1) Elaborate necklaces fashioned from Baltic amber, glass beads, and precious stones—jewelry created in Kaliningrad, Russia. Photographed at TGMS's 2017 expo with permission from Luba Tarasova, owner of Amber Elegance.

(2) Replica of a giant fish, *Xiphactinus audax*, that swam in a seaway dividing North America about 85 mya. Photographed at AAPS's 1992 show, with permission from Mike Triebold/Triebold Paleontology.

(3) Replica of a giant flying fish-eater, *Pteranodon longiceps*, that ranged throughout the Northern Hemisphere about 83 mya. This pterosaur had a wingspan up to 24 feet (7.3 m). Photographed at AAPS's 2002 show; specimen courtesy of Triebold Paleontology.

(4) A rare, highly prized, spiny trilobite, *Comura bultyncki*, of Middle Devonian age, from the Atlas Mountains in Morocco. Photographed at AAPS's 2009 show; specimen courtesy of Brian Eberhardie/Moussa Minerals & Fossils.

Moroccan Sea Creatures: Morocco has been full of surprises. Many of the best specimens of giant marine reptiles and long-spined trilobites have been found in Morocco's colorful mountains and phosphate beds. In the 1990s, buyers could be reasonably confident that spectacular trilobites offered for sale were real. But by 2003, shows in both Tucson and Europe were inundated with fake trilobites from Morocco, mostly fabricated from resin (see Chapter 9).

European Beauties: Europe has always been one of the richest sources of artfully prepared fossils in the world. From Germany's famous Solnhofen limestone and Messel oil shale deposits come a phenomenal array of creatures that range from dragonflies and giant amphibians to flying reptiles and ichthyosaurs. Woolly mammoths, woolly rhinos, cave lions, and cave bears are among the fabulous Ice Age fossils from northern Europe and Siberia that often appear in Tucson. From Italy, we see many exotic fish, crustaceans, and clams; equally stunning specimens have arrived from Lebanon. And from the UK, ammonites and armored fish take center stage.

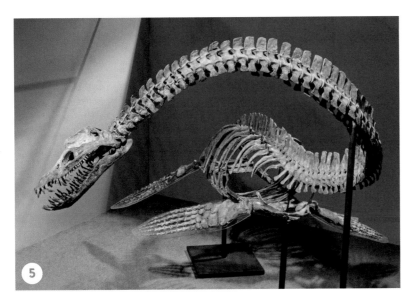

(5) Skeleton of a 93.5 million-year-old marine reptile, *Libonectes atlasense* (= *L. morgani*), a plesiosaur, a predator from Morocco. This specimen, more than 85% original, was on display in the Mineral & Fossil Co-op during the 2015 show. Specimen courtesy of Serge Xerri.

(6) Sea star, *Riedaster reicheli*, of Late Jurassic age, found near Solnhofen, Bavaria, Germany. This specimen—courtesy of Martin Goerlich/Eurofossils—was displayed in at AAPS's 2016 show.

(7) A stingray-like skate, *Cyclobatis major*, of Late Cretaceous age (97.5–91 mya) from Lebanon.

(8) An Eocene (56–33.9 mya) crab, *Harpactocarcinus sp.*, from Italy. Both specimens courtesy of Stefano Piccini/Geoworld Group; Mineral & Fossil Co-op, 2017.

(9) Slightly larger than today's white rhino, the extinct woolly rhinoceros, *Coelodonta antiquitatis*, roamed Europe and Siberia during the last ice age—a species hunted by early man. Photographed in GeoDecor's showroom with permission from owner Tom Lindgren, 2016.

Canadian Curiosities: Canada is well known for its deposits of weird Cambrian life forms, iridescent ammonites, and dinosaurs. Best represented in the Tucson showcase are Frisbee-sized ammonites from Alberta that glow in brilliant hues of red, green, and blue.

Ammonites and Giant Eggs: Glistening clusters of ammonites from Madagascar have hit the Tucson spotlight, and gigantic eggs from Madagascar's extinct elephant bird of Madagascar occasionally appear in fossil showrooms. The first one I photographed was in 1994 (see Chapter 6, *The Miraculous Egg*). From 2008 to 2011, dinosaur nests and eggs from China were a common sight in some of the Tucson fossil shows; but under tightening Chinese export regulations, they have become a rarity.

Top: Shell of a modern chambered nautilus.

Above: Natural cluster of ammonites, nautiloids, and clams of late Early Cretaceous age (113–105 mya) from Madagascar. Maximum diameter of the orange ammonite by right edge of frame is 10 inches (25 cm). Exhibited in Stefano Piccini's/ Geoworld Group showroom in the Mineral & Fossil Co-op; 2018.

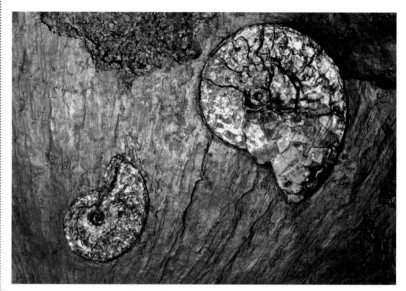

Top: Gem ammonites, *Placenticeras intercalare*, in Late Cretaceous Bearpaw shale (71 mya) from southern Alberta, Canada. Closely packed crystals of aragonite in thin layers diffract white light into spectral colors. Shell fragments are sold as gemstones under the name *ammolite*. The larger shell is 10 inches (25.4 cm) in diameter. Specimen displayed by Canada Fossils, Ltd. in AAPS's 2011 show.

Above: A nest of dinosaur eggs—possibly from a therizinosaur (= segnosaur)—of Cretaceous age (145–66 mya), collected in Henan Province, China. These eggs are smaller than an ostrich egg. Specimen courtesy of geologist Zhouping Guo; displayed at AAPS's 1994 show.

Ancient Life from Down Under: Australians have been bringing opalized shells, exotic sea lilies (crinoids), and specimens of some of the earliest life forms on our planet. And in 2013 they populated the courtyard of Hotel Tucson City Center with life-like models of dinosaurs. These dinosaurs have continued to show up but have declined in numbers since 2013.

Dinosaur Country: The United States has a wealth of fossil deposits and a long history of fossil collecting. By far, the US leads the world for the most dinosaur fossils ever found. They are concentrated in the West along the spine of the Rocky Mountains. Needless to say, dinosaurs are well represented at the Tucson showcase, as are many other American fossils. To highlight just a few: giant fish, trilobites, and sea lilies from the Midwest; dinosaurs, prehistoric mammals and ammonites from the Dakota badlands; early amphibians and mammals from the red beds of Texas; petrified wood from Arizona; shark teeth from the Southeast coast; early marine organisms and seed "ferns" from the Northeast; saber-tooth cats from tar pits in California; and of course the vast array of plants and animals from the Green River Formation of Utah, Wyoming, and Colorado.

People involved in the art and science of fossil collection and preparation represent a relatively cohesive and sharing community. Conflicts can develop over digging sites, but my impression is that most commercial fossil dealers are less motivated by the cash value of their paleo-treasures than by the thrill of developing a deeper understanding of life on Earth.

Above: Lukas Hebdon, age 2, delighted by an original skull of a ceratopsian dinosaur, *Triceratops horridus*, of Cretaceous age (145–66 mya), from the Hell Creek Formation in South Dakota. Displayed by the Black Hills Institute of Geological Research at the 1992 AAPS show.

Top: Prehistoric fish, *Mioplosus sp.*, with roots of a *Sabalites* palm; from the Green River Formation, Wyoming; Eocene age (56–34 mya). Displayed at the 2017 22nd Street Show; specimen courtesy of Thomas Perner/Earth Art Gallery.

Above: This crinoid (= sea "lily"), *Jimbacrinus bostocki*, inspired the creators of creatures for the science fiction film *The Matrix*. From Permian deposits (299–252 mya) in Western Australia. Specimen courtesy of Tom Kapitany/Crystal World/Australia; 2018 AAPS Show.

MESSENGERS FROM THE PAST

Fossils (from French for *something dug up*) are the preserved remains of ancient organisms. Usually what is preserved is a hard part, such as a tooth, shell, or tree trunk.

When buried, the hard parts may be replaced by minerals carried in groundwater or may dissolve away, leaving merely an impression. Sometimes only traces of an animal's behavior are left behind, such as burrows and tracks. These are known as **trace fossils (=ichnofossils)**, which also include fossilized feces (=**coprolites**). In specialized environments, such as frigid places and dry caves, we see unusual forms of fossilization without mineral penetration or replacement. **Paleontologists** (from Greek for *scientists who study ancient beings*) have found whole Ice Age mammoths in the Siberian tundra and partially mummified ground sloths that died in dry North American caves some 15,000 years ago. By definition, all fossils are old—evidence of creatures preserved before written history began.

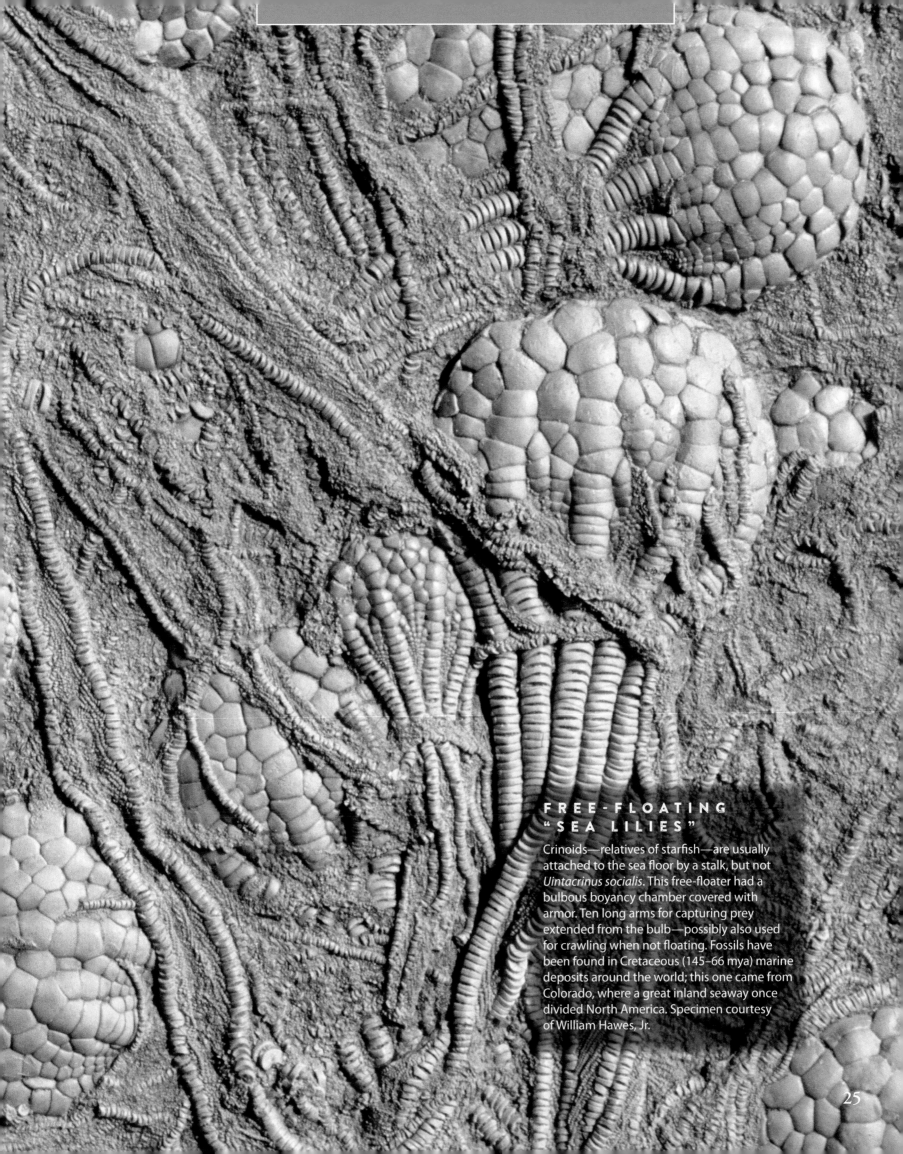

FREE-FLOATING "SEA LILIES"

Crinoids—relatives of starfish—are usually attached to the sea floor by a stalk, but not *Uintacrinus socialis*. This free-floater had a bulbous boyancy chamber covered with armor. Ten long arms for capturing prey extended from the bulb—possibly also used for crawling when not floating. Fossils have been found in Cretaceous (145–66 mya) marine deposits around the world; this one came from Colorado, where a great inland seaway once divided North America. Specimen courtesy of William Hawes, Jr.

25

POSITIVE + NEGATIVE

Split rock nodule, showing positive and negative halves of a *Dactilioceras commune* ammonite from Upper Lias, Yorkshire, England, UK. This fossil dates to the Early Jurassic (201–174 mya). Specimen courtesy of Simon Cohen Fossils.

"SUBZERO"

Skin is rarely preserved with dinosaur remains, so this exceptional fossil was given a nickname: "Subzero." The skin with associated tail vertebrae, chevron bones, and ligaments belonged to duck-billed dinosaur *Brachylophosaurus canadensis*. The mineralized skin clearly shows the pattern, depth, and texture of the scales, which average about 0.4 inch (1 cm) in diameter. Found in Late Cretaceous deposits (94–66 mya) of the Judith River Formation in Montana. Photographed in GeoDecor Showroom with permission from owner Tom Lindgren.

Above: Petrified wood of an extinct tree, *Araucarioxylon arizonicum*, riddled with insect tunnels—trace fossils of the insects that made them. Specimen of Late Triassic age (237–201 mya) from the Chinle Formation in northern Arizona—housed in the collection at Petrified Forest National Park

TRACE FOSSILS

Above: A 2,500-year-old human footprint found buried along a now-dry riverbed that passes through Tucson, Arizona. This is one of many tracks and ancient artifacts discovered by archaeologists in 2016 during the construction of a new bridge. Such tracks in hard, sun-baked mud, later covered by relatively soft layers of silt, can remain preserved as trace fossils. These villagers were among the first farmers of the American Southwest to grow crops in irrigated fields, during what is called the Early Agricultural Period—2100 BCE (=BC) – CE (=AD) 50. Excavated by Archaeology Southwest.

Below: Fossil tracks made by *Dimetrodon*—a distant relative of mammals—on early Permian mud flats of the Abo Formation (299–273 mya), Las Cruces, New Mexico, USA. Specimen courtesy of Jerry MacDonald, Paleozoic Trackways Project sponsored by the Smithsonian Institution and Carnegie Museum of Natural History.

SEALED IN AMBER

Wasp with air bubbles trapped in 15–26 million-year-old tree resin that has hardened into amber. Specimen from the Dominican Republic, courtesy of Jim Work.

RARITY IN GOLD

Top view of an extremely rare specimen, a trilobite, *Triarthrus eatoni*, with its delicate legs, gills, and antennae preserved in pyrite (=fool's gold). At least 99% of all trilobites collected in the world lack evidence of soft tissues. This fossil was found embedded in 450-million-year-old mudstone, in Lewis County, New York. Its body length is 0.8 inch (2 cm). Specimen courtesy of fossil prospector Markus Martin/Gold-bugs.

Optical Illusion: Look at this image multiple times—the trilobite, which bulges from the rock as a positive fossil, will suddenly appear to dip into the surface—the illusion of a negative impression. Look again, and it will switch back to its positive self.

27

Our physical environment is always on the move. Layers of sediment deposited horizontally often get twisted, up-ended, or lost to erosion over time. So geologists must consider earth movement patterns when reconstructing the past and dating fossil beds. This naturally split sandstone boulder on the edge of Lake Powell in southern Utah, USA, clearly shows its original horizontal layering.

MISSING FOSSILS, MISSING LAYERS, MISSING TIME

Each fossil is minor miracle. Right from the start, the odds were against it. For a plant or animal to become a fossil, it must be buried quickly. If not, the remains may be eaten or broken up by scavengers or scattered by floods. Even after remains are "safely" buried, if the chemical environment proves unfavorable or the temperature is too high, the hard parts may decay or dissolve away. Moreover, the sheer weight of overlying sediments or earth movements may crush the organism beyond recognition. **Taphonomy** is the study of an organism's journey from its death to becoming a fossil. If a fossil survives intact to emerge once again at the surface, chances are good that weathering and erosion will destroy it before anyone can find and collect it.

The record of Earth history is not complete. There are gaps in the rock record that geoscientists call **unconformities**. These gaps represent extensive periods of erosion, times when sediments were being stripped away faster than they were being deposited. One cycle of uplift and erosion long ago could have erased rock layers that represented an entire chapter of Earth history.

Sometimes, rock layers that are exposed in one region are missing from nearby areas. In the US state of New Mexico, for instance, Jurassic dinosaurs are found only in its northern half—*Stegosaurus, Allosaurus, Camarasaurus,* and *Diplodocus.* But in the southern part of New Mexico, dinosaurs from this period are missing. Why? Perhaps rocks of Jurassic age have eroded away; perhaps they were never deposited there in the first place; or they may have been covered or destroyed by volcanic activity.

JUMBLE OF DINOSAUR BONES

Scattered skeletal remains of *Camarasaurus* in the "Wall of Bones" at the Quarry Exhibit Hall of Dinosaur National Monument (**above**). Here visitors can see hundreds of dinosaur bones that have been partially exposed for public viewing, including those of *Allosaurus*, *Apatosaurus*, *Diplodocus*, and *Stegosaurus*. The park contains over 800 paleontological sites along the Utah-Colorado border, USA, with interpretive exhibits that tell the story of this prehistoric environment in Late Jurassic times (164–145 mya).

Above right: A much-prized skull of *Camarasaurus* in the park's "Wall of Bones." Dinosaur skulls are often missing; and without the skull, we may never know what the animal ate, how large its brain was, how keen its senses were, and how similar it is to related species.

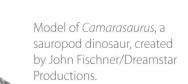

Model of *Camarasaurus*, a sauropod dinosaur, created by John Fischner/Dreamstar Productions.

FISH + POOP = WHAT?

Fossils "as found" can be deceptive. This small fish, *Knightia sp.*, for example, certainly appears to be pooping a mass (probably a croc coprolite) as big as its "maker"—an obvious impossibility! The two were joined in an amusing way during fossilization. But other juxtapositions like this are not so clear and always require cautious interpretation by paleontologists. Specimen from Wyoming's Green River Formation, courtesy of Stefano Piccini/Geoworld Group

PSEUDOFOSSILS

Some of the unusual and beautiful geological creations in the mineral world resemble fossils but are not. The most commonly seen *pseudofossils* are mineral patterns and nodules formed during the weathering of sedimentary rocks. Colorful fernlike growths of manganese dioxide crystals, known as *dendrites*, are often seen in limestone slabs from Germany and Utah, sometimes fringing the edges of real fossils.

Smooth rock nodules have been mistaken for dinosaur eggs. On 12 August 2012, geologists at a leading Russian university announced the discovery of 40 rock objects 10–39 inches (25cm–1m) in diameter that they believed were dinosaur eggs. These spherical surprises turned up in an unusual geological context. A week later, after viewing a video of the find, a Russian paleontologist discredited this report, which had been sensationalized worldwide.

Paleontologists continue to search for the earliest life forms that colonized our planet. Without hard parts, all that can be found are carbon films and *impressions*, some microscopic in size. A few that have been studied extensively are clearly biological in origin. But others might be purely geological. With more questions than answers, challenging discussions erupt among geologists and paleontologists. The scientific community is curious and critical, always demanding new information for a deeper and clearer understanding of the world around us.

DENDRITES

An extinct, predatory ray-finned fish, *Aspidorhynchus acutirostris*, with well-developed dendrites—fern-like patterns of mineral crystals along the fissures and around the fish. This Late Jurassic (164–145 mya) specimen, courtesy of Martin Goerlich, is from the Solnhofen Formation in Bavaria, Germany.

MINERAL GARDENS

People unfamiliar with dendrites often mistake these mineral growths for fossilized plants. They form in water with high a concentration of metals, often iron and manganese. Branching crystals develop in cracks or between layers of rock. This elaborate pattern of dendrites was found in Germany's Solnhofen limestone.

BEAUTIFUL >> BUT BIOLOGICAL?

Presently, experts cannot agree whether these 2.7-billion-year-old layers in *Banded Tiger Iron* from Pilbara, Western Australia, are of biological origin. If formed by microbes—a microbialite (=stromatolite)—this would be one of the oldest examples of life on Earth. Specimen courtesy of Tom Kapitany/Crystal World, Australia.

MINERAL OR MULTICELLULAR?

Some paleontologists claim that these odd geometric discs in "672-million-year-old" shale from Western Australia's Ranford Formation are evidence of soft-bodied multicellular life forms. Others disagree. Mineralogists make the point out that some minerals grow in flat, radial shapes under heat and pressure as water carries them through layers of rock. And some such reactions, including crystal growth, are influenced by bacteria that feed on these chemicals. So reconstructing the distant past will always be a challenge and open to debate. Specimen courtesy of Tom Kapitany of Crystal World, Australia.

LIVING FOSSILS: A DATED CONCEPT

Living fossils are said to be modern organisms that are almost identical to species known from the distant past. The term "living fossil" is not a scientific concept, has no formally accepted definition, and is falling out of fashion as new discoveries in paleontology rise to the surface. Some modern plants and animals do appear remarkably similar to species found in the fossil record, but appearances can be deceiving.

When Charles Darwin coined this term in 1859 in his famous book *On the Origin of Species by Means of Natural Selection*, he made reference to Australia's duck-billed platypus and South American lungfish. Darwin said: "These anomalous forms may almost be called living fossils; they have endured to the present day, from having inhabited a confined area, and from having thus been exposed to less severe competition." Darwin clearly recognized the role of stable and "friendly" environments in slowing the process of evolutionary change. On the flip side, species living in unstable environments must adapt quickly, or face extinction. In all living things, evolution is and always has been a work in progress.

Coelacanths are ancient fish that represent an off-shoot from the ancestors of amphibians and all other land vertebrates. Their fleshy, limb-like fins are supported by bones that allow the paired fins to move in an alternating pattern much like most amphibians, reptiles, and mammals. They appeared in the fossil record about 360 mya, and the group was once widespread in both marine and freshwater environments. One of them, *Latimeria*, has survived in deep water around the Comoro Islands between southern Africa and Madagascar. And in 1997, a second living species was discovered 6,000 miles away, in Indonesia.

Anatomical and molecular studies indicate that coelacanths are not "living fossils." Early descriptions of these fish assumed that their anatomy was "frozen in time," which has since been proven to be incorrect. Their fins are similar, but other anatomical features vary dramatically. They are far more diverse than once thought, represented by more than 70 species in 30 different species groups (**genera**, plural of **genus**). Furthermore, scientists have no fossils for the genus *Latimeria* to compare with the two living species.

Recent genetic studies of *Latimeria* by an international team of 91(!) biologists provided a blueprint for understanding the evolution of four-legged land animals. Their broad-based molecular research compared genes of aquatic and terrestrial species. They focused on immunity, urea/nitrogen excretion, and the development of fins, tail, ear, eye, brain, and the sense of smell. They found that coelacanths are evolving, though slowly, and confirmed that another living group of lobe-finned fish, the **lungfish**, are more closely related to the first vertebrate animals to have crawled onto land.

Above: Coelacanth fossil, 22 inches (56 cm) long, probably an undescribed species. This superb specimen is of Late Jurassic age, found in Solnhofen limestone, Bavaria, Germany. Specimen courtesy of Raimund Albersdoerfer.

Right: Cast of a living coelacanth, *Latimeria chalumnae*, a species of lobe-finned fish native to the West Indian Ocean and known for its vivid blue pigment. This life-sized reconstruction was made from a freshly-captured fish—it is 4 feet (1.2 m) long. Few museum specimens exist, and living coelacanths are Endangered species. Cast courtesy of Mike Triebold/Triebold Paleontology, Inc.

32

Fossil lungfish, *Dipterus valenciennes* (6.7 inches/17 cm in length), found in Old Red Sandstone of mid-Devonian age, in Caithness, Scotland. Today, there are only six living species of lungfish, all in freshwater habitats. These air-breathing fish are believed to be the closest relatives of amphibians. Specimen courtesy of British fossil hunter Chris Moore.

Above: One of four living species of African lungfish, *Protopterus dolloi*. This 1901 illustration was published in *Les Poissons du Basin du Congo* by Belgian-British ichthyologist-herpetologist George Albert Boulenger. During his lifetime, Boulenger described more than 2500 animal species, many from Africa.

LUNGFISH

Like the two known species of living coelacanths, six species of lungfish are descendents from a bygone era. A wide diversity of their lobe-finned relatives swam in Devonian seas 419–359 mya—well represented in the fossil record—and one or more of these early forms spawned terrestrial vertebrates. Living lungfish that biologists can study today are not the direct ancestors of land animals, but biologists have learned a lot from these survivors.

Even within this small sample of six modern lungfish—four in Africa, one in South America, and one in Australia—we see an impressive diversity in anatomy and life-style. Each species is on its own evolutionary path. For example, the Australian lungfish (*Neoceratodus forsteri*), the most primitive of the group, has four fleshy limb-like appendages and only one lung. It relies on gills to extract oxygen from water, if there is enough dissolved oxygen—if not, it must surface to breathe air, using its lung. Australian lungfish cannot live out of water for more than a few hours.

In contrast, the South American lungfish, *Lepidosiren paradoxa*, and the four species of African lungfish are more similar to each other. All have two lungs that have taken over the function of respiration. Their gills are reduced to the point of being useless for oxygen uptake, so without access to air at the surface, they will drown. If their watery home dries up, they can tunnel into mud, secrete a slimy protective cocoon, slow-down their metabolic functions (*estivate*), and wait underground for the next heavy rain.

All six species live in fresh water and are omnivorous—their diet includes frogs, fish, mollusks, insects, and plant matter. The lungs of lungfish are intricately folded into smaller air sacs, which increases the surface area for gas exchange. And like adult amphibians, blood circulation in lungfish is divided—it flows through two pathways, one devoted to respiration and the other directed to the rest of the body—made possible by a amphibian-like heart. The smallest living lungfish, native to East Africa, grows to 2 feet (61 cm) in length; and the largest, the Australian lungfish, commonly reaches twice that size, even up to 4.9 feet (1.5 m). Their four "lobe-like" fins aren't very impressive looking; nevertheless, similarities in skeletal architecture clearly link their limb anatomy to that of amphibians, non-avian reptiles, birds, and mammals.

The Australian lungfish, *Neoceratodus forsteri*—photo © Zoo Leipzig in Germany.

Ginkgos first appeared in the fossil record about 270 mya, and these 170-million-year-old *Ginkgo huttoni* leaves were collected in Yorkshire, England. The oldest known fossils of *Ginkgo biloba*, the only living species, are about 70 million years old.
Specimen courtesy of Mike Baatjer.

GINKGO: A "LIVING FOSSIL"?

The *Ginkgo* is a biological oddity and a tough tree. It is unique among the five living groups of seed plants—the Ginkgophyta, with only one living species, *Ginkgo biloba*. The veins in its leaves radiate out from the base and never interconnect the way they do in other broadleaf plants. Aspects of their reproduction are also surprising. Like mosses, ferns, and cycads, their sperm (packaged within pollen grains) are equipped with tiny whip-like structures for locomotion, a primitive trait among plants. Male ginkgos have pollen cones, and female trees have paired, naked ovules that grow at the end of a stalk, not sheltered within a cone or flower. They rely on tree-to-tree wind pollination.

Ginkgos were once common worldwide when huge dinosaurs roamed the land, but their origins remain a mystery. The oldest known fossils of *Ginkgo biloba* are about 70 million years old, and today its range in the wild is restricted to small refugia in China. These slow-growing trees are believed to live up to 3500 years and are now widely cultivated. Young and old trees alike share an extraordinary resistance to disease, and ginkgos were among the few survivors of the atomic bomb blast on the Japanese city of Hiroshima in 1945. Their trunks were destroyed, but the trees quickly re-sprouted from the roots and have continued to flourish.

Ginkgo biloba is often called a "living fossil" because its external anatomy appears unchanged over the past 70 million years. But on the molecular level, the picture is far more complicated. A team of Chinese scientists has discovered that this plant's genome is more than three times the size of our human genome, and much of its DNA is devoted to building complex proteins that offer resistance to environmental stress. *G. biloba* is armed with an astonishing arsenal of defensive chemicals that repel pests, including insects, fungi, and bacteria. If we could magically transplant a *G. biloba* from 70 million years ago into today's *Ginkgo* habitat, would it survive?

Fresh leaves of *Ginkgo biloba* with a fossilized, 60-million-year-old leaf of a closely related, extinct species, *Ginkgo cranei*, from the Sentinel Butte Formation, Morton County, North Dakota, USA. Microscopically, the leaves and female reproductive organs of these two look-alikes differ structurally. Specimen courtesy of Mark and Karen Haverstein/Lowcountry Geologic.

FOSSILS INSIDE OUT

Animals that appear to be "frozen in time" deserve a closer look. Keep in mind that our on-going reconstruction of the past is a bit like assembling a three-dimensional jigsaw puzzle with most of the pieces yet to be discovered. Modern crocodiles, for example, resemble their ancient croc-like counterparts, but these similarities are superficial. Crocodile evolution is far more complex than was once believed. Many recent fossil finds and molecular analyses of living species have clarified our picture of crocodilian evolution. We now know that they were a very diverse group with dozens of weird and wonderful species that populated land and sea during the Mesozoic Era. Twenty-four species—alligators, crocodiles, caimans, and gharials—have managed to survive; but not by remaining "unchanged." The only constant on Earth is change. And organisms that fail to adapt to a changing environment will perish. So it comes as no surprise that classic examples of "living fossils" are falling by the wayside, one by one, as both extinct and modern species are scrutinized more closely with modern analytical techniques.

Art print made by transfer of acrylic craft paint from a living Sunda gharial at St. Augustine Alligator Farm, Florida— a technique that requires patience and a good understanding of the animal's behavior. All proceeds from the sale of these prints are donated to wildlife conservation projects.

Superficially, this extinct crocodilian, *Steneosaurus bollensis*, looks similar to today's Indian gharial, *Gavialis gangeticus*. But detailed studies of its anatomy tell a different story. This prehistoric croc had heavy body armor on its back and belly, unlike any living species. And studies of the biomechanics of its skull indicate that *S. bollensis* had a quicker, stronger bite than the gharial. Some of its extinct relatives even evolved into forms that looked more like modern killer whales than gharials. The long evolutionary history of crocodylomorphs has been, and will continue to be, dynamic. Specimen from Early Jurassic deposits in Holzmaden, Germany—courtesy of Raimund Albersdoerfer.

FOSSILS
LOST & FOUND

3

Watery environments with accumulations of sediment favor fossil formation, places like the sea floor, lagoons, lakes, river valleys, and floodplains. In general, the finer the sediment, the better the preservation.

But even under favorable conditions for fossilization, the fossil record of life on Earth will always be uneven and skewed. As you would expect, species abundantly represented when alive are more likely to show up in the fossil record than those that were scarce. Similarly, those with hard body parts are more likely to become fossils than are fragile, soft-bodied ones like worms and jellyfish. And organisms living in arid environments unfavorable for fossilization will also be relatively rare.

COASTAL TIDAL FLAT
Lagoons and estuaries with sediments deposited by tides or rivers provide ideal conditions for fossilization. Coastal habitats stabilized by mangrove trees nourish much of our world's seafood supply and support nesting populations of shore birds. They also protect warm seacoasts from storm surge and erosion. Red mangrove on Andros Island, Bahamas.

EXQUISITE PRESERVATION

Well-armored bony fish like this extinct relative of gars (Order Ginglymodi) often preserve well. This specimen, 15 inches (39 cm) in length, is of Late Jurassic age, from Solnhofen limestone in Germany. Specimen courtesy of Raimund Albersdoerfer.

BADLANDS

Arid environments work against fossil deposition, but they are a boon to fossil collectors. Fossils are easy to spot in desert **badlands**, where plants are scarce and surface soils have been stripped away by erosion. The Sioux Indians coined the term *mako sica* (literally "bad land") to distinguish the barren mounds and spires in the Dakotas from the surrounding grassy plains. Today, geologists reserve the term for landscapes of soft sedimentary rock that have been extensively eroded in a dry climate, often by flash floods and seasonal rivers. Without plant roots to help hold the soil in place, erosion happens quickly. In only ten years, as much as 3 inches (7.6 cm) of clay erodes away in Arizona's Painted Desert badlands. At that rate, a hill 25 feet (7.6 m) high would vanish in a thousand years. Speedy erosion of badlands has brought to the surface some of the richest fossil deposits in the world, from the American West and Australia to the Karoo of South Africa and the Gobi Desert of Mongolia.

AGATIZED WOOD

Polished cross-section of petrified wood, *Araucarioxylon arizonicum* (above), a specimen in the collection at Petrified Forest National Park (PEFO), AZ, USA. Petrified logs of Late Triassic age continue to emerge from eroding badlands in PEFO (**left inset**). Panoramic view (**top**) of Painted Desert badlands from Tawa Point in PEFO.

IN CONTRAST

Finding fossils is uncommon in moist, humid terrestrial environments where leaf litter and soil accumulate fast. Fossils are present but are more difficult to locate beneath the forest floor. That's why road-cuts, riverbanks, sea cliffs, and old rock quarries are often rewarding places to discover fossils. Coastal clay cliffs around Lyme Regis, England, for example, are constantly crumbling into the sea, exposing 180-million-year-old ammonite shells and the bones of giant sea-going reptiles. Some of the most famous fossils come from old quarries, including the first dinosaur bone, found in a limestone quarry in the center of London in 1676. And the first known feathered dinosaur, *Archaeopteryx*, was unearthed in a quarry near Solnhofen, Germany in 1861.

FERN

Ferns, cycads, huge amphibians, croc-like phytosaurs, and early dinosaurs are among the fossils found in northern Arizona's desert badlands, providing a snapshot of life during the Triassic Period. Specimen of fern *Phlebopteris smithii* courtesy of Petrified Forest National Park.

MONSTERSAURIA

Note the resemblance between skulls of the Late Cretaceous lizard *Gobiderma* (**above left**) from badlands of Mongolia and that of living Gila monsters, *Heloderma* (**above right**), from the American Southwest. Fossils suggest this ancient group of lizards, the Monstersauria, lived alongside rat-sized early mammals in Eurasia and North America during the reign of giant dinosaurs. Gila monsters specialize in eating bird eggs and infant rodents, a life-style that might date back to the Cretaceous! Skulls courtesy of Mark Norell, Division of Paleontology, & C. J. Cole, Dept. of Herpetology/AMNH.

WOOLLY MAMMOTH

Right: Skeleton of adult male woolly mammoth of Pleistocene age from Siberia, Russia. The maximum span across these impressive tusks is 77 inches (2 m). Photographed in GeoDecor Showroom/Tucson Mineral & Fossil Co-op with permission from Tom Lindgren.

Below: Authentic replica of Lyuba, a rare and beautifully preserved woolly mammoth calf found frozen on the bank of the Yuribei River in northwestern Siberia. To fabricate this model, Dutch artist Remie Bakker used measurements and photos of the original mummy to sculpt and cast it. On permanent exhibit at the Mammoth Site & Museum in Hot Springs, South Dakota, USA.

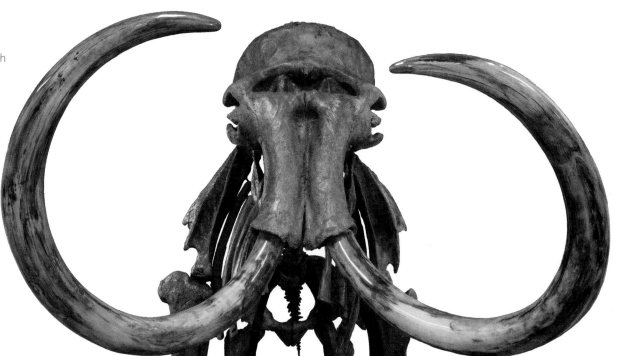

MUMMIFICATION

If environmental conditions prevent the decomposition of dead tissues, an organism's remains can be naturally preserved for millions of years. Bones of Ice Age bears, ground sloths, giant flightless birds, and prehistoric humans, for example, have been found in dry caves, sometimes with mummified skin, hair, connective tissue, stomach contents, and dung.

One of the clearest windows to the past has come from animals entombed in ice or permafrost—soil that never thaws out. As our climate continues to warm and polar ice melts, skeletons and carcasses that have been frozen for more than 10,000 years are being discovered. These finds are often incredible, complete with internal organs, blood, and even DNA fragments—fertile ground for research into the life and times of prehistoric animals.

In 2007, a nomadic reindeer herder discovered an exquisitely preserved mammoth mummy in northwestern Siberia. This month-old female calf named Lyuba died about 41,000 years ago. Her body has been under intense study in Russia, Japan, and the USA. CT scans have revealed that the calf was in good health when she died; and large amounts of mud found within her trunk, mouth, and trachea suggest that she may have suffocated in soft mud. Lyuba's permanent home is Russia's Yamal-Nenets Regional Museum.

Reconstructed skeleton of *Ursus spelaeus* —Latin for "bear cave"—given this name because most fossils of this species have been found in caves. Cave bears lived in Europe and Russia during the Pleistocene and became extinct during its coldest glacial cycle about 20,000 years ago. This fossil was found in Germany—specimen courtesy of L&J Fossils/Austria.

Dry caves often provide ideal conditions for the preservation of skeletons, dung, and mummified animal remains. Cueva del Candelaria in Guatemala.

DEATH TRAPS

About 26,000 years ago, a sinkhole that filled with artesian spring near what is now Hot Springs, South Dakota, became a death trap for Pleistocene mammals. When animals came to drink, some could not escape from this steep-sided waterhole, and for about 700 years their bones accumulated in a bed of mud on the floor of this 60-foot-deep sinkhole that later dried up. In 1974 a developer discovered this treasure-trove of fossils while excavating for a housing project. Scientists soon realized that lying beneath their feet was the largest concentration of Columbian mammoth remains in the world! **The Mammoth Site** continues to be an active research dig, now housed in an impressive building with exhibits and on-site paleontologists working with volunteers in these bone beds. By 2019, 58 Columbian mammoths and three woolly mammoths had been unearthed, along with giant short-faced bear, camel, llama, wolf, and other Ice Age mammals.

Flooded limestone caves and sinkholes (known as cenotes in Latin America) often yield exquisitely preserved fossils. Cave divers in Madagascar have found thousands of bones representing relatively recent extinctions. These "young" remains possess tissues not fully mineralized, which paleontologists call **subfossils**. The species mix includes 10-foot-tall (3m) flightless elephant birds, horned crocodiles, and 5,000-year-old gorilla-sized lemurs.

SUPER SHARK

Enya Kim is dwarfed by the reconstructed jaws of *Carcharodon megalodon*, commonly called "megalodon," the largest extinct megatooth shark known to have lived on our planet. The teeth are real, but the cartilaginous jaws had to be fabricated. Photographed in GeoDecor Showroom/ Tucson Mineral & Fossil Co-op with permission from Tom Lindgren.

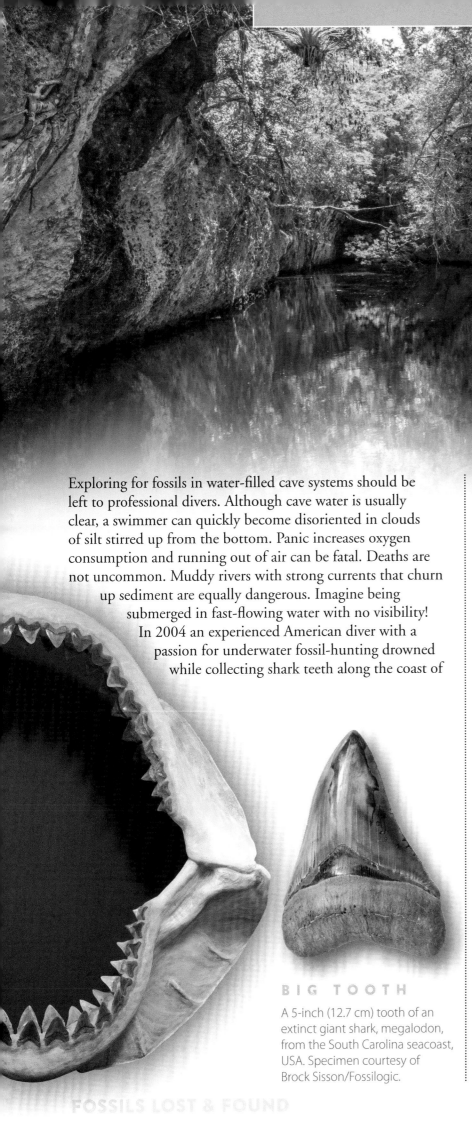

CENOTES & BLUE HOLES

Adventuresome divers are often attracted to water-filled sinkholes, which can be inland or coastal and interconnected by caves. In Spanish they are called "cenotes," and in English, "blue holes." Divers often risk their lives exploring these underwater cave systems to search for fossils. This cenote was photographed in Zapata Swamp, Cuba.

Exploring for fossils in water-filled cave systems should be left to professional divers. Although cave water is usually clear, a swimmer can quickly become disoriented in clouds of silt stirred up from the bottom. Panic increases oxygen consumption and running out of air can be fatal. Deaths are not uncommon. Muddy rivers with strong currents that churn up sediment are equally dangerous. Imagine being submerged in fast-flowing water with no visibility! In 2004 an experienced American diver with a passion for underwater fossil-hunting drowned while collecting shark teeth along the coast of Georgia. The current was fast, the water pitch black, and he was diving in a deep hole full of debris. Vito Bertucci had devoted his life to finding enough teeth of the right size to reconstruct the jaws of the world's largest known shark, **megalodon**, meaning "big tooth," (**center image**). Based on the size of their teeth and vertebrae, these prehistoric giants were thought to be, on average, about 35 feet (10.7 m) in length. A great white shark could have swum through its massive jaws. Scientists know they roamed oceans worldwide throughout the Miocene (23–5.3 mya) and had vanished before the onset of the last Ice Age (2.6 mya).

DANGEROUS DIVING

A diver ventures into silt-laden seawater—never a good idea—at the entry to an underwater cave on Andros Island, Bahamas.

BIG TOOTH

A 5-inch (12.7 cm) tooth of an extinct giant shark, megalodon, from the South Carolina seacoast, USA. Specimen courtesy of Brock Sisson/Fossilogic.

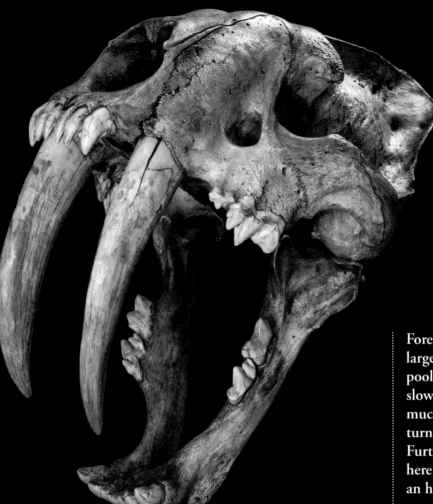

SABER-TOOTH "TIGER"

Common names can be misleading—saber-tooth cats are prehistoric members of the cat family, but none are "tigers." *Smilodon fatalis*, shown here, is the California State Fossil. This superbly preserved skull was retrieved from La Brea tar, along with more than 2,000 others. Specimen courtesy of Fossilogic.

TAR PITS

Quite literally, some prehistoric creatures got stuck in time—in tar or tree resin. In a few well-known places crude oil has seeped to the surface through cracks in Earth's crust, to become sticky puddles of tar. Natural deposits of this kind in California, Trinidad, Venezuela, Peru, and Azerbaijan have yielded thousands of fossilized plants and animals. All known **tar pits** were formed during the last Ice Age, the Pleistocene. Large mammals were common victims of tar pits, but why? Weren't they smart enough to avoid this foul-smelling goo? In some places pools of tar were hidden beneath grass. And where tar is clearly exposed, a predator could easily end up mired with its meal during the heat of a chase. A sheet of tar less than 2 inches (5 cm) thick, is sticky enough to entrap a large mammal.

The **Rancho La Brea Tar Pit**, located in the heart of Los Angeles, has yielded an incredibly rich and well-preserved collection of Pleistocene fossils. The Natural History Museum of Los Angeles County now houses more than 3.5 million La Brea specimens, representing more than 660 species of mammals, birds, insects, mollusks, and plants—an entire ecosystem preserved in time. But there is an odd imbalance in these fossils: about 90% of the mammals and most of the birds are predators or scavengers, like saber-tooth cats, dire wolves, vultures, and eagles. How could this be, when prey animals normally far outnumber the meat-eaters?

Forensic studies of bone-munching insects in the remains of large herbivorous mammals that got stuck in these shallow pools of tar (like camels and horses), suggest that they sank slowly into the ooze over a period of 17–20 weeks. This left much of the meat visible, tempting to carnivores that in turn got stuck and died in the company of their last meal. Furthermore, these remains tell us that animals were active here for at least 30,000 years. So if you were to witness an herbivore and its associated carnivores getting fatally entrapped in this goo, when, on average, might you expect to see this again? Not for another 10 years, according to biologists who have inventoried the bones found in this pit.

TRAPPED IN TAR

Fossilized bones of a Pleistocene elk from California's McKittrick Tar Seep. Preparation, as found, courtesy of John Alcorn.

PORTHOLES TO THE PAST

Wounded or infected cone-bearing trees like pines secrete a thick, sticky resin from their bark to form a protective shield while healing, much like a scab over a wound in humans. Small animals and bits of plant matter get stuck in this goo, which, when buried, heated, and compressed turns to a semi-precious gemstone called **amber**. Tar preserves only the hard parts of animals, such as bones and the exoskeletons of insects. But resin entombs victims quickly and seals them off from outside elements. This protects delicate appendages usually lost during fossilization. But, to be clear, the premise of Hollywood's Jurassic Park films—that of rescuing 100-million-year-old dinosaur DNA from a mosquito— remains in the realm of fiction.

The best-known deposits of amber are located in countries around the Baltic Sea and in the Dominican Republic, Mexico, and Myanmar (formerly Burma). They are 15– 99 million years old. The majority of the supply is sold in gemstone markets. Scientists are most interested in rare pieces that contain fossilized plant or animal matter (organic inclusions)—about 90% of which are insect remains, with most of the rest being spiders. Even more rare are millipedes, scorpions, frogs, lizards, mammal hair, flowers, leaves, moss, lichens, and feathers. Exciting finds have been coming from Myanmar lately, which include a snake and even the tail of a tiny dinosaur-like bird, presumed to be a juvenile, with a total length of only 1.4 inches (3.5 cm).

AMBER MINING

Working by candlelight in a hot, cramped mountainside tunnel, Dominican miners chisel their way through bedrock to find precious deposits of amber.

DOMINICAN AMBER

Above: Amber from the Dominican Republic is prized for its clarity and rich inclusions of plant and animal fossils. Shown here are a small lizard and a centipede. Specimens courtesy of El Museo del Amber Dominicano and collector Jim Work.

Right: Miner pauses at the entrance to a pick-and-shovel amber mine in the Cordillera Septentrional, Dominican Republic. These drift mines in broken, fragmented rock have no reinforcement and are extremely dangerous.

FOSSILS LOST & FOUND

GEOLOGICAL
TIME & DRIFTING
CONTINENTS

4

Nearly 200 years ago, paleontologists discovered that fossils occur in an orderly sequence. Older fossils were in the lower sedimentary layers, the first to be deposited, and the younger fossils were on top, in the more recent deposits.

Knowing this allowed geoscientists to determine the **relative ages** of fossil-bearing rocks, which quickly led to the development of a worldwide **geologic time scale**, a calendar showing the history of life on Earth (see illustration on next page spread). Later dating techniques, such as measuring the rate of decay of radioactive elements, allowed geologists to assign **absolute ages** to rocks without fossils. Mineral-rich volcanic rocks are especially useful in absolute dating. For archaeological sites and very recent deposits containing trees or freshwater clam shells, scientists could also measure time by counting yearly growth rings. And as discussed in greater detail later in this chapter, dating techniques continue to be expanded and refined.

Sunset Crater Volcano erupted in Arizona around 900 years ago. By combining multiple dating techniques—from tree-ring samples, paleomagnetism, and radioisotope data (see section on *Measuring Time*)— a scientific team recently determined that the eruption lasted no more than a few months, sometime between AD 1085–1090 (= CE 1085-1090). This young cinder cone is now protected as a US National Monument.

Crack in a crust of pahoehoe lava from Kilauea Volcano, Hawaii. To date fossils, it's often necessary to date rocks associated with them. Deposits of volcanic rock and ash are especially valuable—they contain radioactive elements that begin to change/decay at a known rate after being ejected. This allows scientists to date the deposits accurately (see text).

EON	ERA	MILLION YEARS AGO	PERIOD		EPOCH	
P H A N E R O Z O I C	**C E N O Z O I C** (Greek for recent life) / Age of Mammals	0	QUATERNARY		HOLOCENE	
					PLEISTOCENE	
		2.6	TERTIARY	NEOGENE	PLIOCENE	
					MIOCENE	
				PALEOGENE	OLIGOCENE	
					EOCENE	
					PALEOCENE	
	M E Z O Z O I C (Greek for middle life) / Age of Reptiles	66.0	CRETACEOUS		LATE/UPPER	
					EARLY/LOWER	
		145	JURASSIC		LATE/UPPER	
					MIDDLE	
					EARLY/LOWER	
		201	TRIASSIC		LATE/UPPER	
					MIDDLE	
					EARLY/LOWER	
	P A L E O Z O I C (Greek for ancient life) / Age of Amphibians	252	PERMIAN		LOPINGIAN	
					GUADALUPIAN	
					CISURALIAN	
		299	CARBONIFER-OUS	PENNSYLVANIAN	LATE/UPPER	
					MIDDLE	
					EARLY/LOWER	
				MISSISSIPPIAN	LATE/UPPER	
					MIDDLE	
					EARLY/LOWER	
	Age of Fishes	359	DEVONIAN		LATE/UPPER	
					MIDDLE	
					EARLY/LOWER	
		419	SILURIAN		PRIDOLI	
					LUDLOW	
					WENLOCK	
					LLANDOVERY	
	Age of Marine Invertebrates	444	ORDOVICIAN		LATE/UPPER	
					MIDDLE	
					EARLY/LOWER	
		485	CAMBRIAN		FURONGIAN	
					MIAOLINGIAN	
					EPOCH 2	
					TERRENEUVIAN	
		541				

PROTEROZOIC EON: 541-2500 million years ago

ARCHEAN EON: 2.5-4.0 billion years ago (earliest signs of life).

HADEAN EON: 4.0-4.6 billion years ago (formation of the Earth's crust)

PRECAMBRIAN TIME

Geological Time Scale (updated by The Geological Society of America in August 2018). The most recent updates can be found online— visit the International Commission on Stratigraphy, where chart translations are available in multiple languages.

HISTORY OF LIFE

Pending sixth mass extinction from unchecked spread of human populations & unsustainable use of Earth's natural resources

All modern life forms. Expanding human civilizations with written language, agriculture, & evolving technologies.

11,700 ya

Giant land mammals coexist with archaic human hunters, *Homo erectus* & Neanderthals. First *Homo sapiens*.

2.58 mya

Rise of mammoths, ground sloths, camels, rhinos, saber-tooth cats, 2-ton armadillos, & more. Continued expansion of grasslands & grazing mammals.

5.3 mya

Modern families of birds established. Time of giant sharks & crocodiles. First anthropoid apes & signs of early humans.

23 mya

Global expansion of grasslands. Reign of many strange mammals, now extinct, like oreodonts, entelodonts, & giant hornless rhinos. Marine mammals diversify.

33.9 mya

First ancestors of many modern mammals, including horses & whales. Oceans warm & teeming with bony fish & other sea life. Crocodiles, turtles, & snakes flourish.

56 mya

Reptilian giants of the Cretaceous now extinct, but crocodiles & sharks survived. Fearsome birds evolve; first penguins. Explosive rise of land mammals. Early primates.

66 mya

66.0 mya: famous K-T (Cretaceous-Tertiary) mass extinction. 75% of all species vanished; likely cause, an asteroid collides with Earth.

Flowering plants spread rapidly and dominate the land, with help from insects. Reef-building rudist clams, starfish, sea urchins, & planktonic diatoms thrive. Monstrous marine fish & reptiles abound, including the first mosasaurs, along with their favorite prey, ammonites. Rapid diversification of ray-finned fish. Dinosaurs and flying reptiles rule the land and air, in the company of birds.

Trees dominating the land were conifers, ginkgos, & especially cycads; first flowering plants. Most modern insect families appear in fossil record. Dinosaurs & pterosaurs flourishing. First snakes. First birds & small placental mammals. Sea populated with ichthyosaurs, giant crocodiles, modern-looking sharks & rays, ammonites, & belemites.

201 million years ago: fourth major mass extinction; about 80% of all species lost.

A time of change and rejuvenation after the largest mass-extinction on Earth. Land well populated with coniferous trees and cycads. Insects excel. Ocean rich with microscopic phytoplankton, ammonites in coiled shells, dolphin-like ichthyosaurs, & long-necked plesiosaurs. Frogs, salamanders, lizards, sphenodontids, crocodilians, & turtles established. Rise of small dinosaurs, flying reptiles, & rodent-sized mammals.

252 mya: third major mass extinction, the largest ever. Vast majority of marine life lost, along with about 70% of all terrestrial species.

Large trees: seed-ferns, conifers, ginkgos, & cycads. Increasing insect diversity. Abundant amphibians, some up to 30 feet long! Primitive mammal-like reptiles, e.g. sail-backed Dimetrodon (a top predator of the Permian) & shrew-like cynodonts, a group from which all modern mammals evolved.

Shallow, warm-water lagoons occupied by single-celled planktonic animals, bottom-dwelling filter-feeders, sharks, & bony fish. First octopus & freshwater clams. Great coal-forming swamp forests of ferns, club mosses, and horsetails. Earliest cone-bearing plants. First land snails & flying insects. Spread of amphibians. First reptile, a lizard-like creature. First shelled eggs suited for life on dry land, e.g. eggs with the embryo protected by fluid-filled amniotic membranes, which set the stage for the evolution of reptiles, birds, & mammals.

360-380 million years ago: pulses of mass extinction events; about 70-75% of all marine species lost.

Sharks, armored fish, ammonites, & sea scorpions colonize the ocean, with reef-building coralline algae, horn corals, & brachiopods. Lobe-finned fish give rise to first vertebrate with amphibian features.
First seed plants and forests with woody trees; scale trees, rushes, horsetails, & ferns populate the land.

First spiders, scorpions, & centipedes.
Proliferation of early land plants; first vascular plants.
Oldest coral reefs, with widespread diversification of sea lilies (crinoids), brachiopods, trilobites, & mollusks. Fish with jaws appear.

444 million years ago: first major mass extinction; about 85% of all aquatic species lost.

First corals & diverse echinoderms—sea stars, crinoids, blastoids, & cystoids. Many kinds of trilobites, brachiopods, mollusks (nautiloids, clams, & snails), & jawless fish (some freshwater). Reefs dominated by algae & sponges. Colonization of land by early plants.

All life forms were marine; land was largely covered by shallow inland seas. Proliferation of invertebrate sea animals, including strange soft-bodied organisms, jellyfish, sponges, mollusks, brachiopods, crustaceans, & trilobites.
First animals with backbones, aka vertebrates: jawless fish. Several extinction events happened in the second half of the Cambrian.

Few fossils. First simple forms of life: bacteria, aquatic algae, stromatolites, jellyfish and sponges.
Earliest signs of life: single-celled organisms without nuclei–about 3.5 billion years old.

Earth's crust begins to form about 4.55 billion years ago.

ILLUSTRATION BY PAUL MIROCHA

The calendar of life in rock is divided into **eons**, which in turn are subdivided into **eras**, **periods**, and **epochs**. As our illustration shows, the first three of the four recognized eons of geologic time are jointly referred to as the **Precambrian**, a time when the first traces of life on Earth appear in the fossil record. This happened early in the Archaean Eon, 3–4 billion years ago! Younger rocks from the Proterozoic Eon contain many primitive life forms—the fossil remains of unicellular life, like bacteria, as well as the first multicellular animals that depended on oxygen.

The rest of the history of life on Earth, up to the present day, falls within the **Phanerozoic Eon**, which begins with the **Paleozoic Era** and the **Cambrian Period**. What is often called the Cambrian "explosion" is marked by the sudden appearance of more diverse and complex life forms in the fossil record. Cambrian seas teemed with animals of many shapes, sizes, and life styles, some free-swimming and others confined to the sea floor. Most basic body plans of all major animal groups on Planet Earth had been established in the fossil record by the end of this period, and a few plants and animals began venturing onto land.

Note that for graphic clarity, our time interval rectangles are of equal size in this chart even though they represent widely varying time spans. The time scale will tell you when one time interval ends and the next begins, a scale determined by significant changes in the fossil record, such as major extinction events. As new information becomes available, these dates are refined and published online by the International Commission on Stratigraphy.

To understand and map the earth history of an area, geologists also identify and name rock **formations**. A formation is a body of rock or rock particles with characteristics that distinguish it from all neighboring rocks. These characteristics may include rock type, its particular mix of minerals, or a unique assemblage of fossils. Most formations have been named after a geographic feature (such as the Green River Formation in the American West) or a place (Solnhofen, a town in Bavaria, Germany). The Lias Formation in England, France, and Germany evidently got its name from the provincial English pronunciation of the word *layers*, in reference to its evenly stratified limestone.

To reconstruct the evolutionary history of life on Earth, we need to know *when* each fossilized creature died. Without a time machine, determining the exact age of a fossil can be challenging. And the farther we go back into the time scale of geological events (*deep time*), the harder it becomes to calculate these ages accurately. Scientists can assign ages to fossils by using two main dating methods: relative (= indirect) dating and absolute (= direct) dating.

When geology officially became a science in the 1700s, **relative dating** was the only method available for establishing the chronological ages of rocks and fossils. Geologists recorded the positions of sedimentary rock layers relative to each other. Even though precise ages could not be assigned to these rocks or the fossils within them, relative dating has remained a useful starting point. It gives paleontologists a way to organize life forms in a meaningful sequence of appearance, proliferation, and disappearance of species within the fossil record. Species that are well studied, abundant, and widespread from a short, specific period of time are known as **index fossils** (= *key fossils*). Knowing the age of an index fossil can sometimes offer a relative age estimate for other fossils.

LAMP-SHELLS AS INDEX FOSSILS

Brachiopods, commonly called *lamp-shells*, are marine animals that superficially resemble clams, but their anatomy is so unique, they are classified in their own phylum. They were abundant in oceans worldwide during the Paleozoic Era and are among the most common fossils of that age. Their shell features are variable and easy to study, so brachiopods are often used as index fossils. Some species were about the size of a big pinhead; others grew to 12 inches (30 cm) in width. Certain lineages appeared suddenly in the fossil record and others vanished quickly, making them useful for correlating layers of sedimentary rock in different places. Brachiopods have helped to define boundaries between Paleozoic periods in the calendar of life. These "butterfly shells" of *Mucrospirifer* from Sylvania, Ohio, USA, are mineralized with pyrite. Most brachiopods in this species group have been found in rock layers of Middle Devonian age, 393–383 mya, in Eurasia, northern Africa, South America, Canada, and the USA. Specimen courtesy of Jon Kramer/Potomac Museum Group.

A breakthrough that led to **absolute dating** came in 1898 when French physicists Marie and Pierre Curie discovered the phenomenon of radioactive decay, work that earned them a Nobel Prize in physics. Radioactive elements occur naturally throughout the universe, and the Curies found that radioactive atoms are inherently unstable. Over time, radioactive isotopes (= "parent atoms") decay into stable new isotopes (= "daughter atoms"). In 1907 American radiochemist Bertram Boltwood found that isotopes in rocks can be used to measure their age. For example, uranium is a parent isotope that slowly decays into lead, its daughter isotope. These decay rates are fixed and constant. And the amount of time required for *half* of the original parent sample to decay into the daughter isotope is known as its **half-life**. The ratio between old and young isotopes acts like a sort of geological clock that tells us the time when a rock or fossil was formed.

The half-life clock of decaying isotopes follows a predictable curve and not a straight line. Think of it as trimming a grove of trees: if it takes 5,000 years to trim half of them, trimming the other half will require an additional 5,000 years, and so forth. So it would take 10,000 years to trim ¾ of the grove, and 15,000 years to trim 7/8 of the grove.

All organisms live in a shower of cosmic rays that pass through Earth's atmosphere. In fact, every person is bombarded by about half a million cosmic rays every hour. These collisions with atoms in living tissues—which are packed with carbon-based molecules—generate the radioactive isotope Carbon-14. C-14 has a relatively short half-life of 5,730 years, so **radiocarbon dating** can only be used reliably to date fossils up to about 42,000 years— at this point, all of the parent atoms have decayed.

Many other radioactive isotopes with longer half-lives can be used to date volcanic rocks associated with fossils that are millions, even billions of years old. Although often called "radiocarbon" dating, this is better known as **radiometric or radioisotope dating**. Potassium-Argon (decay of K-40 to Ar-40) is popular among dinosaur paleontologists because it can be used to accurately date layers of volcanic ash in bone beds of Mesozoic age. To improve confidence in the results, scientists often analyze and compare decay rates of several different elements to see if they all agree.

Explosive volcanic eruptions are characteristically short-lived, and every layer of ejected ash and rock (*tephra*) has a unique geochemical signature. So these deposits are relatively easy to recognize and measure using radiometric dating. Fossils found within, above, or below a layer of volcanic debris can then be assigned relative ages based on their position. This is called **tephrochronology**.

VOLCANIC ASH DEPOSITS

In deep time, explosive volcanic eruptions happened quickly—much like this recent one in 1980 at Mount St. Helens, Washington, USA. Eruptions of this type throw a blanket of ash and rocky debris over the landscape, a layer that can be precisely dated with radioisotopes. Fossils found within or near these deposits can be assigned an accurate relative age by association. Photo by Michael Doukas /Cascades Volcano Observatory / USGS.

PETRIFIED SEQUOIA

Close-up detail of growth rings in a cut-and-polished slab of a fossilized 38-million-year-old sequoia tree. It was cut from a log weighing 28,000 pounds, the largest piece of petrified wood ever found. The whole slab—89 inches (2.3 m) wide x 68 inches (1.7 m) tall—has 816 annual growth rings. Modern sequoia trees commonly reach ages of 2000–3000 years. The log was excavated from volcanic basalt in central Oregon, USA. Other plants found in the same deposit indicate that the climate was shifting from tropical to a cooler, drier, more seasonal environment. Clear growth rings develop only in trees that experience well-defined growing seasons. Specimen courtesy of Ralph Thompson/Russell-Zuhl Petrified Wood.

In 1929, Arizona astronomer Andrew Douglass discovered that tree rings record time, the beginning of the science of **dendrochronology**. Growth rings are visible when a tree trunk is cut, and in environments with one annual growing season, you can expect to see one growth ring added per year of life. So by counting the rings in a living or freshly cut tree, the tree's age can be estimated with great accuracy. Growth rings in old wood (like timbers in prehistoric dwellings) can often tell us how many years a tree lived, but rarely *when* it lived. So scientists are now combining radioisotope dating with tree-ring counts to pin down age estimates. Furthermore, trees growing in the tropics do not add growth rings in a predictable way because there is no well-defined annual growing season. Arizona's ancient petrified wood, for example, was deposited about 215 mya when Arizona had a tropical climate. So paleontologists must rely on radioisotopes with a long half-life, which usually requires dating volcanic soils that contain the petrified wood.

Studies of changes in our planet's magnetic field, **paleomagnetism**, are often combined with radioisotope dating to increase accuracy. Convection currents in Earth's liquid core produce a strong magnetic field, so our whole planet behaves like a giant magnet. Because some minerals respond to these magnetic forces, they leave a permanent record of the direction and intensity of this field in ancient rocks. Rocks formed in the last 780,000 years "point" to the magnetic north pole, just like your compass needle does. But 785,000 years ago, your compass needle would have pointed south, not north! Such periodic *magnetic reversals* in polarity are predictable and show up as "fingerprints" in rocks formed during the past 200 million years. By knowing when these switches have happened through geological history, paleontologists have been able to create a time scale that can be correlated with radiometric dates across the fossil record. Of special interest, no mass extinctions have been in sync with magnetic reversals.

Scientists continue to add new methods to their arsenal for dating, always striving for better accuracy. This growing list of dating options includes luminescence, electron spin resonance (ESR), fission track dating, amino acid racemization (described later), obsidian hydration, and cosmogenic nuclide exposure—enough to make your head spin! Explaining all these techniques is well beyond the scope of this book, but two of them are often mentioned in *paleoarchaeology*, studies of human evolution.

Luminescence is a group of dating techniques used to measure electrical energy stored by some minerals when exposed to heat or light—quartz rocks, for example, or calcite-containing fossils like teeth and corals. Mineral grains collected from a field sample are analyzed in a laboratory. When the sample is stimulated by heat or UV radiation, the amount of trapped energy released as light can be measured. Luminescence is often used to determine when these minerals were last exposed to sunlight—which resets the electrical charge to zero—before being buried. These techniques can be applied to samples from 1,000 to 500,000 years old.

From the biological side come other techniques for paleontological dating. One of these is based on a measureable change in amino acid molecules—the building blocks of proteins—after a living creature dies. Amino acids in living tissues have what's called a "**L**eft-handed" molecular structure; but at the time of death these molecules begin to undergo a structural transformation into a "**R**ight-handed" form—a phenomenon called **racemization**. If environmental conditions remain stable, this conversion proceeds at a constant rate. So knowing this rate of conversion, researchers can determine relative age by measuring the percentage of **L**-forms that have changed to **R**-forms in a sample since the animal's death. Under ideal conditions, this can allow fossil dating up to 100,000 BP (years before present). But its usefulness depends on a thorough understanding of the fossil's environment, from the time the animal died to the present day. Racemization is a temperature-sensitive chemical reaction, so an organism buried in a cold cave will change at a slower rate than one in sands of the Sahara Desert.

For more than 50 years, evolutionary biologists have been studying genetic changes at the subcellular level that behave like a ticking **molecular clock**, the name given to another new dating technique. They measure rates at which specific DNA and amino acid sequences mutate and get "shuffled" over time. By knowing the rate of change between species, biologists can estimate the timing of important evolutionary events, such as when a genetic lineage of animals split (diverged) into two groups or acquired distinctive anatomical features. To reconstruct these chronologies, scientists must often calibrate their findings against dates established for well-known fossils.

In summary, every dating technique has its strengths and weaknesses, and reaching an accurate estimate of a fossil's age usually requires a combination of techniques. To test their reliability, dates are confirmed using at least two different methods, often involving multiple independent labs to cross-check the results. When only one method is possible, researchers are less confident of the results. The International Commission on Stratigraphy publishes updated geologic time scales every few years, based on the newest data for major time lines. Older dates may shift by a few million years up or down, but younger dates tend to remain stable. For example, in the 1960s a date of 65 mya was set for the end of the Mesozoic Era, the Age of Reptiles. Since then, this date has been tested and re-tested with more modern techniques and equipment, and the refined date of 66 mya now remains uncontested and accurate to within a few thousand years. That's a confidence level of about 99%!

A   1.4 Billion Years Ago

B   900 Million Years Ago

E   100 Million Years Ago

F   60 Million Years Ago

MOVING CONTINENTS

"The earth, instead of appearing as an inert statue, is a living, mobile thing." ~ Canadian geophysicist J. Tuzo Wilson, 1968

Belgian geographer-mapmaker Abraham Ortelius published the first atlas of the world in 1570. His volume of maps titled *Theatrum Orbis Terrarum* (*Theater of the World*) compiled the work of numerous scientists, complete with drawings of sea monsters. Ortelius noted that the coastlines of the continents appear to fit together like pieces of a puzzle. And after giving this considerable thought over the next 26 years, he proposed that continents in the eastern and western hemispheres must have been joined at one time . . . then torn apart by "earthquakes and flood."

In 1910 German astronomer-meteorologist-explorer Alfred Wegener (1880–1930) started to dig deeper and hypothesized that about 300 million years ago Earth's landmasses were united into a giant supercontinent. He named this hypothetical supercontinent **Pangea** (= Pangaea), from Greek meaning *all the Earth*. Thus began his 20-year quest for evidence to support his theory of **continental drift**. Wegener wrote several books to expand and refine his ideas, published in 1915, 1920, 1922, and 1929.

Wegener noted that geologists had found distinctive rock strata in South Africa that were identical to those found in Brazil. Geologists had also found matching scrape patterns from ancient glaciers that had moved along the coasts of eastern South America and western Africa. He considered mountains too—proposing that as continents moved, their leading edge would encounter resistance, compress, and fold upward to form mountain ranges. He suggested that India, for example, had drifted northward to "plow into" continental Asia and form the Himalayas. Furthermore, drawing from literature in paleontology, Wegener pointed out that identical fossilized plants and animals from the same time period were found in South America and Africa (see map of mesosaur fossils on the next page spread). The same was true for fossils found in Europe and North America, and in Madagascar and India.

When Wegener formally presented his ideas between 1912 and 1915, established geologists in England and America viewed him as a kook. In those days geologists favored the notion that Earth's crust was fixed and could never move. To explain similar fossils on widely separated continents, they proposed land bridges that had later sunk into the ocean. They lacked any solid evidence

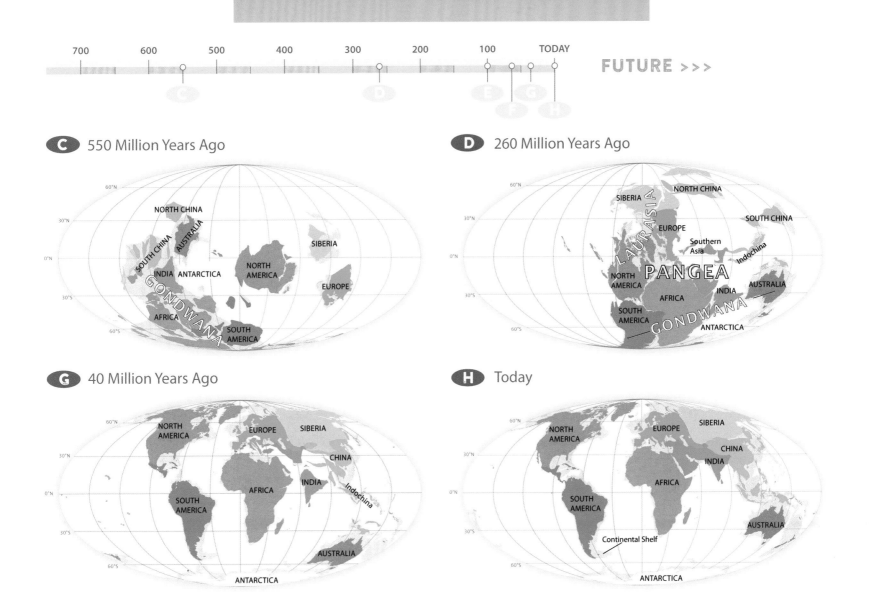

C 550 Million Years Ago

D 260 Million Years Ago

G 40 Million Years Ago

H Today

to support their hypothesis; nevertheless, most frowned upon ideas coming from this "outsider," Wegener, an accomplished meteorologist with a PhD in astronomy. No one could imagine a mechanism for moving continents to new places on the globe.

Although Wegener never thought of Earth's crust as being fixed in place, he was unable to come up with a reasonable explanation for how continents could "drift" through a solid ocean floor. Groping for ideas, he suggested centrifugal force from our spinning planet. Geologists countered that Wegener failed to understand the nature of rocks. Unfortunately, Wegener died at an early age during an Arctic expedition in 1930, but a few English-speaking geologists who saw wisdom in his work continued to pursue his theory of drifting continents.

In 1928 a brilliant young British geologist, Arthur Holmes, speculated that **convection currents** in Earth's sub-surface **mantle** might be the mystery force driving continental drift. To understand convection, just watch the circulation of liquids in a pot of boiling soup. Holmes proposed that rocks heated by the decay of radioactive minerals in our planet's interior will flow towards the surface in a solid state. And with repeated cycles

of circular heating below and cooling above, rocks will spread slowly and carry continental "rafts" with them. Holmes, by the way, designed the first geological time scale using radiometric dating *before* earning his PhD in 1913! Nevertheless, for decades most established geologists ignored his convection theory.

ON THE MOVE

Above: Earth's landmasses have always been on the move. Pieces of our planet's crust have drifted together to form supercontinents, have split apart, and have later reunited to form new supercontinents. Geoscientists have identified several major cycles of supercontinent formation and fracture over the past 3.5 billion years, the most famous being Pangea. *Pangea* formed early in the Permian Period about 299 million years ago, remained as an assembled landmass for about 100 million years, and began to break up during the Jurassic Period about 175 million years ago. *Rodinia* was a giant supercontinent that preceded Pangea, lasting from about 1100 mya to 700 mya. And continents as seen today are by no means in fixed positions. Geophysical models predict that in 250 million years North America, Eurasia, Africa, and Australia will merge to form a new supercontinent in the Northern Hemisphere—nicknamed *"Amasia."*

Global reconstruction of continental movements through deep time were customized for this book in 2020 by Paleogeographer Andrew Merdith, University Claude Bernard Lyon 1, France.

CLUES FOR CONTINENTAL DRIFT

Mesosaurs were small semi-aquatic, fresh-water reptiles that lived early in the Permian Period, about 299–270 mya and then vanished from the fossil record. At that time, South America and Africa were joined at the hip, part of the supercontinent Pangea. Fossilized mesosaur remains (family Mesosauridae) have been found at sites marked with orange dots on this map, one example of the distribution of fossils through time that supports the theory of continental drift. Data source: 37 collection records by multiple authors and mapped on *Fossilworks.org*, a web-based portal to an enormous Paleobiology Database—a non-governmental, non-profit public resource, *PaleoBioDB.org*— well worth a visit! Data are contributed by hundreds of paleontologists from around the world.

Stereosternum tumidum, a mesosaur fossil from Permian deposits in the Irati Formation, State of Sao Paulo, Brazil. The skull of this specimen is 2.4 inches (6 cm) long. Specimen courtesy of the Black Hills Institute of Geological Research.

TECTONIC PLATES

World map of today's major and some minor tectonic plates— about 40% of the world's human population lives within plate boundary zones. Earthquakes, seismic sea waves (= tsunamis), or volcanic eruptions may be triggered where plates meet and move. Geophysicists have learned a great deal about tectonic behavior and can better predict catastrophic events, but we will always be at the mercy of our restless planet. Map customized for this book in 2020 by Paleogeographer Andrew Merdith, University Claude Bernard Lyon 1, France.

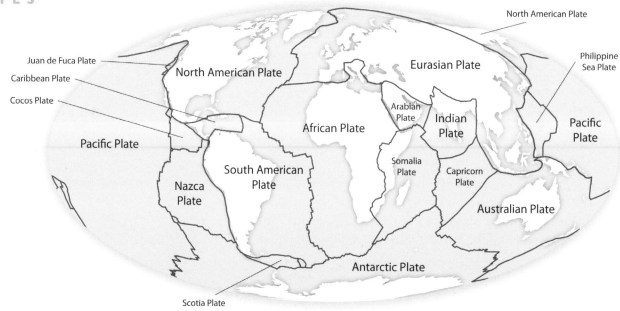

Independent studies of the sea floor in the 1960s supported Holmes's ideas and led to the discovery of **plate tectonics**. While in the Navy, American geologist Harry Hess surveyed the topography of the ocean floor using the ship's sonar equipment. By charting patterns of sound waves bounced off the sea floor, Hess made groundbreaking discoveries on the origin of ocean basins, island arcs, mountains, and the movement of continents. Hess proposed that hot magma rising from under the oceanic crust in global rift zones—long cracks in the Earth's crust between tectonic plates—would expand and push oceanic plates apart as it cooled.

Interdisciplinary studies of the sea floor using submersible vehicles, sediment analyses, earthquake detection equipment, modern dating techniques, and imagery from outer space have all helped to crystallize our modern view of plate tectonics. Today, scientists recognize 11 major tectonic plates and several minor ones; and they are always on the move. Where plates collide, one is usually forced below the other, down into the mantle, a process known as **subduction**. Most plates move

1–2 inches/year (2–5 cm/yr), about the same speed that your fingernails grow. In other words, during a human lifetime of 72 years, continents would move, on average, about 9 feet (2.7 m). For our scientific understanding of the world, plate tectonics has become as important in geology as the theory of evolution is in biology.

GLOSSOPTERIS LEAVES

Some plants are also excellent indicators of continental drift—*Glossopteris* seed ferns are among them. These woody shrubs and trees evolved and flourished in the Permian (299–252 mya) and went extinct at the end of that period. Their fossilized remains are sprinkled across southern parts of South America and Africa; Australia; New Zealand; India; and Antarctica—a puzzling geographic distribution that only makes sense on a map of Gondwana. Back then, all of these landmasses were in close contact—see previous page spread. This Permian specimen of *Glossopteris browniana* leaves—courtesy of Terry Manning—is from Dunedoo, New South Wales, Australia.

COLLECTING FOSSILS

5

Fossils have caught the eye of collectors for at least 3,300 years. Around 1300 BCE (= BC) ancient Egyptians stumbled across pitch-black bones of some strange creatures.

They gathered nearly three tons of these mysterious bones. Many were treated as sacred objects, wrapped in linen and placed in tombs. During the 1920s British explorers Guy Brunton and Sir Flinders Petrie shipped crates of these fossils to the Natural History Museum in London, where they sat unopened and neglected until 2016, according to historian Adrienne Mayor. A research grant has sparked new interest in them, so hopefully we will soon know more about the origin, species, and dates of fossils in this collection.

THE BACK OF BEYOND

Vehicle sheltering in shade of a giant boulder at a fossil collecting site on the remote western edge of the Sahara Desert, in Maider region of Morocco. Some of the world's richest fossil deposits lie in such inhospitable places. Primitive armored fish of Devonian age (419–359 mya) have been collected here. Photo by paleontologist Serge Xerri of Rabat, Morocco.

**PLASTER
FIELD JACKET**

Skull of a mosasaur cradled in
a plaster field jacket for safe
transport from its collecting
site, a phosphate mine in
Oued Zem, Morocco.
Photo by Serge Xerri.

59

HUNTING FOSSILS

Who are the people behind great fossil discoveries? Most are amateurs: farmers, ranchers, nomads, construction workers, beachcombers, scuba divers, and others who spend a lot of time outdoors. Countless rock hounds and fossil collectors are out there exploring too. Some build careers around collecting, preparing, and selling fossils to museums and other collectors. In fact, most of the specimens on public display in museums were acquired from outside individuals and organizations. Few museums and academic institutions have the budgets or staff to send their paleontologists on collecting expeditions for long periods of time. Armed with academic degrees or not, professional fossil collectors are curious, passionate scientists who are always digging for a deeper understanding of our planet. And many are skilled in the art of fossil preparation (see Chapter 9, *Art of Fossil Preparation & Display*).

Attitudes about fossils vary greatly from country to country. In poverty-stricken countries, fossil collecting and preparation have become important economic drivers in some localities. In Morocco, for example, when village people were starting to find, collect, prepare, and sell fossils, they made tools from what was available, such as old pistons for hammers and nails for chisels. Concerned collectors from Europe began donating modern tools and training enthusiastic citizens in the art and science of fossil collection and preservation. For the advancement of science and the local economy, the outcome has been a win-win.

BOOST TO A LOCAL ECONOMY

Many fossils would never be found or extracted without assistance from local people who know the land and how to survive under harsh conditions. Fossils that are abundant in many countries can provide income for low-wage workers. For example, this group of Moroccans is preparing common fossils to supply a world market with specimens of little interest to paleontologists or other serious collectors. Photo by Serge Xerri.

METICULOUS WORK

Trained in the art of fossil preparation, Malika Tlimes removes matrix from tail bones of a giant aquatic reptile, a mosasaur, collected and prepared in Morocco. Photo by Serge Xerri.

GREEN RIVER FOSSIL QUARRIES

Above: From left to right, Candi Mondi, Corey Trometer, and Joseph Smith lift a freshly excavated slab of limestone in a Wyoming fossil quarry owned by Rick Hebdon (in background) of Warfield Fossils. Quarry workers can avoid summer heat by working at night. Breaking layers apart is labor-intensive—done by hammering long flat chisels between rock layers and then prying them apart with flat bladed shovels.

Right: Green River Stone Company's Quarry Manager Matt Helm and his three helpers—David Dilworth, Dallin Ahrnsbrak, and Jacob Benner—mark the location of fossils exposed in the "18-inch layer" of limestone. This sedimentary layer contains the most fossils, the legacy of an ancient lakebed in what's now southwestern Wyoming. After identifying the location of visible fossils, the limestone is cut and chiseled loose in slabs that can be transported to a lab for preparation (see Chapter 9).

An Early Permian four-legged creature, *Seymouria sp.*, with features of both amphibians and reptiles; from the Red Beds of Texas, USA. *Seymouria* was of special interest to A. S. Romer. Cast on display at Wyoming Dinosaur Center, Thermopolis, WY, USA.

Policies governing the export of fossils between countries can be notoriously perplexing and frustrating to serious collectors. Alfred S. Romer's melodramatic experience in Argentina, reported in a 1967 Harvard Alumni Bulletin, is a classic example, as relevant today as it was back then. Under a 1964 agreement between Argentina's national government and Harvard's Museum of Comparative Zoology (MCZ), this esteemed early American paleontologist (1894-1973) devoted four months to exploring vast, virtually uninhabited deserts of western Argentina for fossils. The research was financed by a grant from the US National Science Foundation, with field assistance and logistic support from the Argentinian Museum of the University of La Plata. Romer and his colleagues were seeking early reptiles in Triassic and Permian deposits that might shed some light on the ancient connection that Wegener proposed between South America and Africa. Geologists had mapped this region but had found no fossils.

After two months of fruitless searching, the team discovered a spectacular bed of fossils. They began the arduous but joyful task of excavating, cataloging, and preparing the specimens for safe transport to the USA, another two months of work. Then, in a surprise turn of events, their truckload of precious cargo was seized on the road by local police. According to Romer, a few local people, jealous of this expedition's success, convinced the governor of the province where the fossils had been collected to confiscate the fossils and forbid their export. A national scandal erupted over this incident. Ten months later, word arrived that the seven crates of bones would be shipped to Boston. But in the midst of some rejoicing among scientists, the Argentinian government was overthrown and Harvard allies in the La Plata museum lost their jobs, followed by six more months of anxiety. Finally, after two years of false hopes and despair, friends in La Plata broke through bureaucratic entanglements. The collection was released from custody and arrived safely on the doorstep of the MCZ.

Problems like this have plagued almost everyone interested in fossil collecting, especially in Argentina, Brazil, Mongolia, and China. Laws in these countries have often been vague or enforced by whim, and documents defining permitting procedures have been difficult to pin down. Ministry offices, local officials, and customs agents often provide conflicting information. Some fossil dealers in these countries issue bogus export certificates. And to make matters worse, corrupt customs agents sometimes sign export forms for a fee, giving the impression that the shipment is legal.

Below: Accurate replica of a *Psittacosaurus* dinosaur nest with five babies, as found. These ceratopsian dinosaur skeletons of Early Cretaceous age were collected in Liaoning Province, China. This plate is 33.5 inches (85 cm) long. Reconstruction by the Black Hills Institute of Geological Research (they own the original as well).

In a widely advertised New York auction, a tyrannosaurid dinosaur skeleton from Mongolia went up for sale on May 20, 2012. Responding to a request from the president of Mongolia, US officials confiscated the skeleton before the winning bidder could claim it. For the first time, Mongolia made it clear that without a special permit from the government, all fossils imported to the United States since 1924 had been "smuggled" out of Asia. This news sent a shockwave through the American fossil community. Before then, the international market had been flooded with dinosaur eggs and skeletons from Mongolia, along with fossil mammals and bird-like dinosaurs from China. Fish, pterosaur, and insect fossils were also coming in from Brazil. So at this point, would these countries try to repatriate hundreds of specimens that had been bought and sold to private collectors and museums around the world?

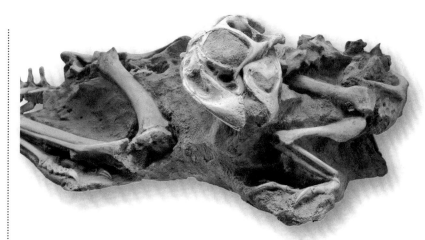

Cast of a partially exposed skeleton of a bird-like dinosaur, *Conchoraptor gracilis*, of Late Cretaceous age (76 mya), from the Gobi Desert in Mongolia. This small dinosaur had a powerful beak without teeth, possibly used to crack the shells of mollusks or seeds. CT scans (see Chapter 8) of another specimen suggest that *Conchoraptor* had a keen sense of vision, balance, and coordination. The skull of this juvenile is 4.5 inches (11.4 cm) in length; cast by Gaston Design.

In Brazil, the National Mining Agency, a division of their Ministry of Mines and Energy, regulates fossil collecting. Under Brazilian law, no one—including Brazilian citizens—is allowed to collect, trade, sell, or possess fossils. But navigating the bureaucracy to get a collecting permit can be daunting, and requests to export fossils for study are likely to be denied. Foreigners are advised to team up with a Brazilian institution and plan to do all of their collecting and research in-country. Unfortunately, policies like this can be counter-productive when outside institutions can offer safer storage and more modern technology for studying specimens.

Brazilian limestone quarries contain some of the world's most important fossil deposits, including crocodile-like mesosaurs, flying reptiles (pterosaurs), and beautifully preserved assemblages of insects and plants. Frozen in time, flowering plants (angiosperms) of Early Cretaceous age were just beginning to diversify, along with the rise of insects as their pollinators 108 million years ago— an important window to the evolution of life on Earth.

The way limestone quarries are managed in Brazil is frustrating both for international scientists and for underprivileged Brazilians. Knowing the value of fossils in these quarry pits, workers sometimes "rescue" and sell what they find, but by doing so, they are breaking the law. Brazilian citizens can buy paving stones from these quarries, but if they contain fossils, owning them, selling them, or exporting them is a crime. Nevertheless, fossils can be seen in walkways and on the sides of buildings where pavers are used in public works projects. Such laws squash curiosity, hurt the local economy, and drive the trade deeper into the realm of bribes and backroom deals.

On 2 September 2018 a fire gutted Brazil's 200-year-old National Museum, home to the country's prized collection of some 20 million archaeological, biological, and geological treasures. In just six hours Brazil's irreplaceable specimens, artifacts, books, documents, and audio recordings went up in flames. Fossils cannot withstand the intense heat of a huge blaze or the weight of a collapsing building. So nearly all of the museum's invaluable paleontology collection and exhibits were lost—dozens of pterosaurs and dinosaurs, thousands of precious insect and plant fossils, two giant ground sloths, the oldest

GONE IN A FLASH

Below: One night in 2018, a fire consumed Brazil's National Museum, along with most of its biological, archeological, and scholarly treasures. This is a grim reminder that all museums are vulnerable. Defective infrastructure, natural disasters, and acts of war can destroy hundreds of years of our collective history. The author created this image by juxtaposing two photographs: flames+smoke from a bonfire blended with a ghosted daytime view of the National Museum before the fire.

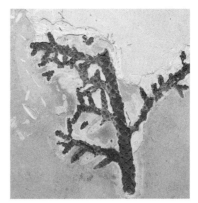

PRIZED FOSSILS FROM BRAZIL

Above: These astounding specimens from Brazil's limestone quarries in the Crato and Santana Formations are among those that escaped the fire that ravaged the National Museum in Rio de Janeiro. They are in the hands of institutions and private collectors outside of Brazil. From left to right: dragonfly; cicada; spider; mayfly; insect wing (probably grasshopper); cockroach; snakefly; and foliage of an ancient conifer (family Araucaceae). Specimens courtesy of Annesuse Raquet-Schwickert and Merv Feick/Indiana9 Fossils.

Update: On 15 June 2020, fire hit another Brazilian museum––the Natural History Museum & Botanical Garden in Belo Horizonte. Fortunately, firefighers were able to contain this one, but not before it damaged five rooms full of biological specimens and Indigenous artifacts.

known armadillo—the list goes on and on. Priceless legacies of the biggest names in Brazilian paleontology were lost in a flash. Among the 2,000 salvaged objects, many of them rock samples, was Luzia, an 11,500-year-old human skull, one of the oldest found in the Americas.

Many scientists considered Brazil's National Museum the best and largest natural history museum in Latin America. The exhibits were beautiful to behold, but the infrastructure was in a state of decay. An improperly installed air conditioning system started the fire. The aging building had no sprinklers, and fire hydrants near the museum failed when firefighters needed them. Ironically, three months before the fire, the Brazilian government had approved $5 million in funding for updates to the museum, including a fire-suppression system, but the funds had not yet been released. It's also worth noting that between 2010 and 2015, fires had devastated four other Brazilian museums, plus their Antarctic research station.

Brazil's tragic loss—and others before it—should serve as a wake-up call to the world's museum community. After San Francisco's earthquake of 1906, the California Academy of Sciences burned down, taking with it 25,000 bird specimens, most of its entomology and herpetology holdings, and its entire library. At the time, this museum had housed one of the largest natural history collections in the USA. And in 2016, a fire ravaged India's National Museum of Natural History. Simply put, no museum is immune to catastrophic events, including floods, earthquakes, terrorism, and war. Many collections are housed in aging buildings, with shrinking budgets. To safeguard their contents, ". . . 'tis the part of a wise man to keep himself today for tomorrow, and not venture all his eggs in one basket," in the words of Miguel de Cervantes (Don Quixote, 1615).

Though costly and time consuming, using modern technology to digitize collections is another forward-thinking strategy. A 2010 study by the American Association of Museums suggested that to remain relevant and financially solvent in today's rapidly changing world, museums should create innovative ways to attract visitors, to forge new research partnerships with other institutions, and to "look outside traditional training programs for bright, interested people and then invest in their continued education."

Even within one country, laws often differ between states or provinces, as in the USA and Canada. Some look kindly on free enterprise; others do not. Perhaps the only universal rule is that no collecting without a special permit is allowed in world heritage sites, national parklands, and other reserves. The rarity and scientific value of certain types of fossils also restricts collecting. And being allowed to collect and keep a fossil does not always give the owner the right to sell or export it. Furthermore, laws are always changing! So read the legal documents—and be careful of hearsay and internet posts. For a brief overview of what to consider before collecting a fossil, check these guidelines, written by a Canadian fossil enthusiast with a legal degree: https://fossilsinsideout.com/fossil-collecting-laws.

Unfortunately, there is no "master list" of laws governing fossil collecting. A United Nations Database of National Cultural Heritage Laws, which includes fossils in their Natural Heritage category, is more focused on cultural antiquities. It is the responsibility of each member state to keep their country's listing current. Much of the information is outdated or incomplete, and for the most part, documents are available only in the member's native language. Consequently, paleontologists have found this database frustrating to use.

The Association of Applied Paleontological Sciences has begun developing a much friendlier format to help collectors understand UNESCO's database. Their country-specific list of *International Laws Pertaining to the Collection, Import, Export, and Sale of Fossil Material* is off to an impressive start. Each summary is backed by links to pertinent laws, contact information for collecting permits, notes, and posting dates. Although still in its infancy, the information is available in AAPS's online Journal of Paleontological Sciences. This project is well worth supporting!

Above: Small fish that once lived in a prehistoric subtropical, freshwater lake—sediments of the Green River Formation—an easy catch for amateur fossil collectors. Among them are this perch-like fish, *Priscacara serrata* (left), and *Gosiutichthys parvus grande*, a herring relative (right). Specimens courtesy of Rick Hebdon and the Black Hills Institute of Geological Research.

PROSPECTING FOR FOSSILS

World-wide, a growing number of companies offer tourists an opportunity to "collect-your-own" fossils on private property. Here, visitors enjoy hunting fossils in Wyoming at a quarry owned by Warfield Fossils— an educational vacation experience. Colorado resident John Fraundorfer and his two daughters, Hannah and Julia, split limestone slabs to find fossils.

COLLECTING RESTRICTIONS

In Green River quarries that allow public fossil collecting, you cannot keep everything that you find. Quarry owners retain specimens of special scientific interest, and all fossils found in US national parks/monuments are protected by law. Specimens shown on this page—all from the Green River Formation in Wyoming—are rare to uncommon.

Bottom, from Left to Right:
Pair of freshwater stingrays, *Heliobatis radians*, from Warfield Fossils, Thayne, WY. These specimens are about 16 inches (41 cm) long.

Partially prepared skull of a small, unidentified mammal in the extinct family Hyaenodontidae—a group of carnivores that lived in North America, Eurasia, and Africa for about 26.1 million years, from the Eocene to mid-Miocene. Maximum length of this skull is 2.2 inches (5.5 cm); specimen collected by Rick Hebdon/ Warfield Fossils.

Skeleton of an Early Eocene tinamou-like wading bird, *Pseudocrypturus cercanaxius*, a member of the extinct family Lithornithidae. This cast is on display in the Visitor Center at Fossil Butte National Monument, Wyoming.

Superb specimen of a young soft-shelled turtle, *Axestemys byssinus*. Width of its shell is 3.5 inches (9 cm). On display at Wyoming Dinosaur Center, Thermopolis, WY.

GREEN RIVER MONITOR

Occasionally prize fossils turn up in Wyoming's Green River Formation—this 5-foot-long (1.5-m) monitor lizard (*Saniwa ensidens*), for example. Monitors (family Varanidae) are confined to the Old World today; they are abundant in Australia, Asia, and Africa. Gila monsters and beaded lizards (family Helodermatidae) are among their closest living relatives in North America. Specimen courtesy of Anthony & Elizabeth Lindgren.

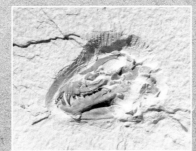

MODERN TREE OF LIFE

This friendly tree of life was reproduced with permission from British vertebrate paleontologist/paleoartist Mark P. Witton. It was featured in his superb 2020 book *Life through the Ages II: Twenty-first Century Visions of Prehistory* (Indiana University Press). These groups of organisms are discussed in the upcoming *Gallery of Life*/Chapter 6, so refer back to this illustration if needed to help visualize relationships among them.

Notice that fungi appear on a side branch of the tree trunk leading to animals, and not on the branch leading to plants. Fungi don't preserve well, so little of what we know about them has been found in the fossil record. Biologists must rely on biochemical and genetic clues to determine who is related to whom. As it turns out, fungi share with many animals a complex structural carbohydrate called *chitin*, which is absent in plants. It appears in the cell walls of fungi and in external skeletons of crustaceans and insects, in mollusk mouthparts, and in the scales of fish and primitive amphibians. Don't confuse chitin with *keratin*, a fibrous structural protein found in skin, hair, nails, feathers, horns, hooves, and reptile scales.

You might also notice the absence of some familiar organisms in this tree of life—algae, amoebae, slime molds, diatoms, etc.—commonly classified in the kingdom Protista. Simply put, biologists cannot agree where these miscellaneous groups of organisms belong, although they undoubtedly originated early in the history of life. As mentioned in the caption related to Ernst Haeckel's *Paleontological Tree of Vertebrates* (see page 4), Protista is a "wastebasket kingdom," an artificial group of mostly single-celled organisms with few shared characteristics. If it's not a plant, not an animal, not a fungus, and not a bacterium, biologists have dumped it into the Protista.

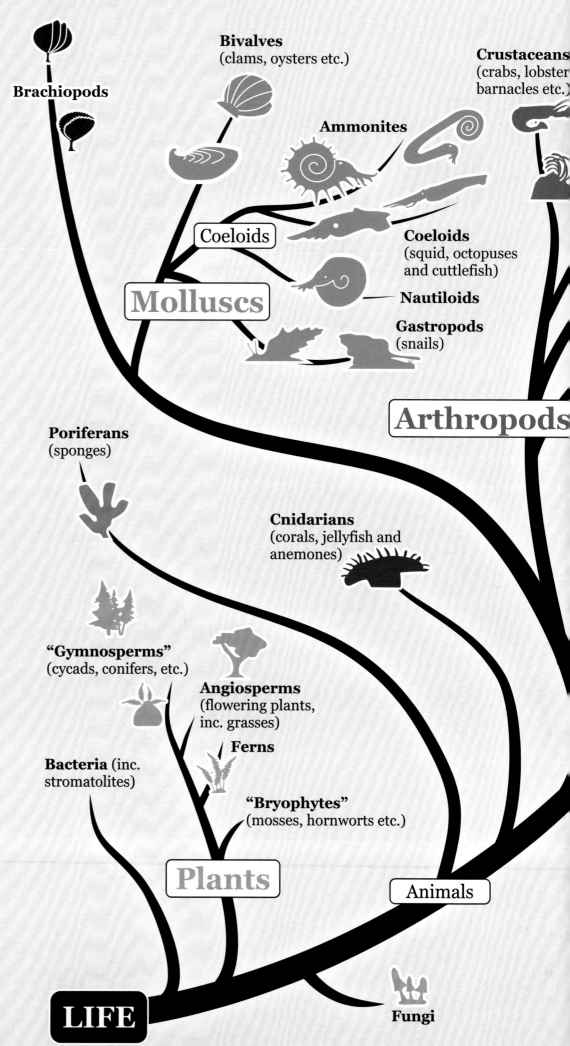

Brachiopods

Bivalves
(clams, oysters etc.)

Crustaceans
(crabs, lobster
barnacles etc.)

Ammonites

Coeloids

Coeloids
(squid, octopuses
and cuttlefish)

Molluscs

Nautiloids

Gastropods
(snails)

Arthropods

Poriferans
(sponges)

Cnidarians
(corals, jellyfish and
anemones)

"Gymnosperms"
(cycads, conifers, etc.)

Angiosperms
(flowering plants,
inc. grasses)

Ferns

Bacteria (inc.
stromatolites)

"Bryophytes"
(mosses, hornworts etc.)

Plants

Animals

LIFE

Fungi

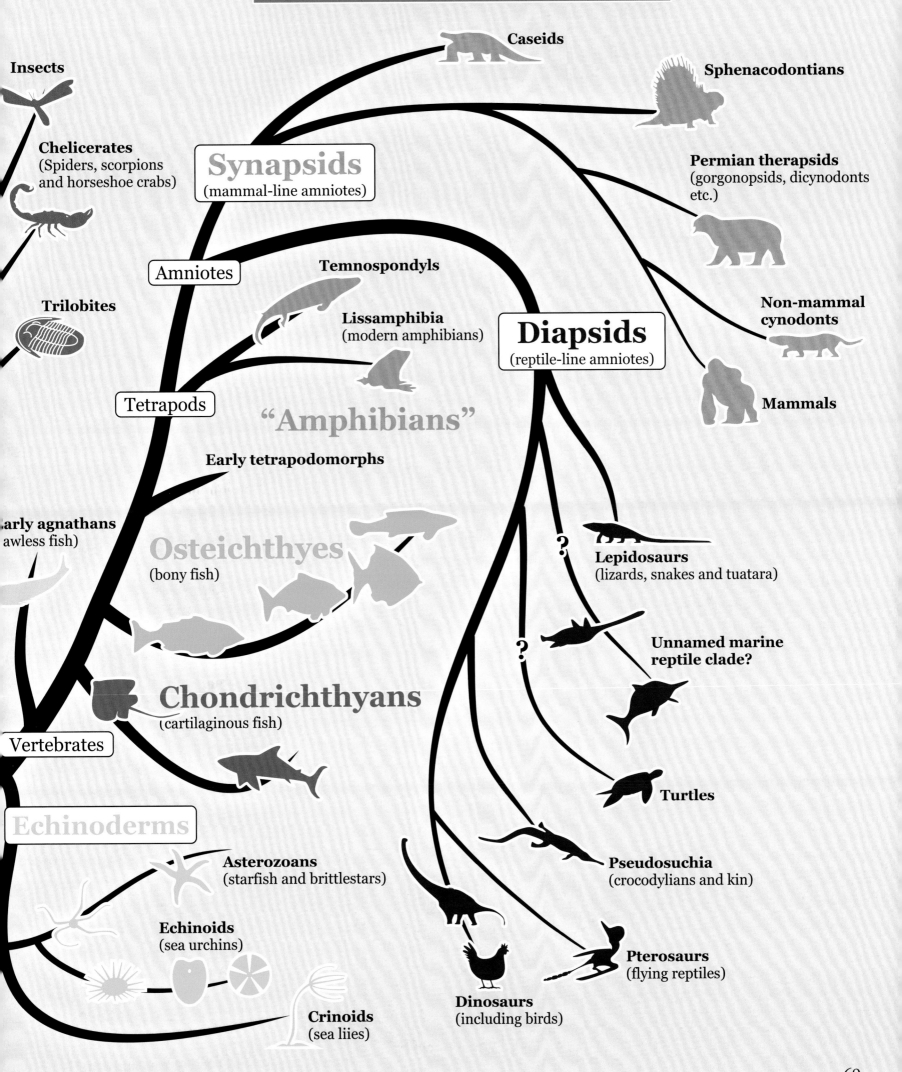

Insects

Chelicerates
(Spiders, scorpions
and horseshoe crabs)

Synapsids
(mammal-line amniotes)

Caseids

Sphenacodontians

Permian therapsids
(gorgonopsids, dicynodonts
etc.)

Amniotes

Temnospondyls

Trilobites

Lissamphibia
(modern amphibians)

Diapsids
(reptile-line amniotes)

Non-mammal
cynodonts

Tetrapods

"Amphibians"

Mammals

Early tetrapodomorphs

Early agnathans
(jawless fish)

Osteichthyes
(bony fish)

Lepidosaurs
(lizards, snakes and tuatara)

?

?

Unnamed marine
reptile clade?

Chondrichthyans
(cartilaginous fish)

Vertebrates

Turtles

Echinoderms

Asterozoans
(starfish and brittlestars)

Pseudosuchia
(crocodylians and kin)

Echinoids
(sea urchins)

Pterosaurs
(flying reptiles)

Dinosaurs
(including birds)

Crinoids
(sea liies)

GALLERY OF LIFE

6

The earliest inhabitants of our planet were tiny—microscopic organisms that, for the most part, are nearly impossible to detect in the fossil record.

"Unaltered" crustal rocks found in Canada, Greenland, and Australia suggest that continental landmasses began forming 4.3–3.5 billion years ago (bya). At the same time, or not far behind, life was brewing in the ocean—in calm coastal waters and in geothermal cauldrons.

PELE'S CAULDRON

In traditional Hawaiian belief, *Pele*—goddess of volcanoes and fire—created the Hawaiian Islands. Here, fresh lava from Hawaii's Kilauea Volcano meets the sea. Pele is both a creator and destroyer, and she has spawned life in unlikely places. Deep within the "midnight zone" on the ocean floor, life thrives around volcanic vents in a bath of hydrogen sulfide.

EARLIEST LIFE

In shallow lagoons rich in calcium and bicarbonate, trillions of individual cells bound together by their filamentous secretions, known as **biofilms**, settled on the sea floor. Interacting chemically with their physical environment, these microbial mats of mostly photosynthetic cyanobacteria (blue-green "algae") trapped minerals and sediments. Over time, these **stromatolites** grew, layer upon layer, to form columns or domes. Although stromatolites were prolific worldwide from 2.5-0.54 bya, they declined rapidly thereafter. But a few modern populations have survived along continental seacoasts and even in freshwater lakes. Those in Western Australia, the Bahamas, and British Columbia's Pavilion Lake are the most famous. Living stromatolites have provided unique opportunities to better understand and interpret their fossilized remains.

Other evidence suggests that life began in high-temperature environments no less than 3.5 bya, quite possibly under conditions similar to what's found in geothermal areas being studied on Earth today. When astronauts were beginning to explore outer space in the 1960s, scientists were hunting for unusual life forms in extreme environments closer to home. American microbiologist Thomas Brock found a host of bacteria and other microorganisms thriving in boiling hot springs of Yellowstone National Park. And soon thereafter, similar discoveries were announced in Iceland, New Zealand, Italy, and other places with active geothermal features on land. These simple, heat-loving organisms, **thermophiles**, require hot water for survival—up to 284°F (140°C). Thermophiles have specialized enzymes that function best under extremely high temperatures. And many thrive in chemical environments that would kill most other organisms—those that are highly acidic or rich in sulfur, for example.

With steadily growing interest in the theory of plate tectonics in the 1960s and 1970s, oceanographers and engineers developed new ways to explore underwater mountain ranges and deep-sea rifts in Earth's crust. Keep in mind that any human or machine sent to the deep sea floor—into the *midnight zone* below 3,280 feet (1,000 meters)—would need protection from frigid temperatures and crushing pressures. No modern military submarine can operate under these extremes. In 1960 the first self-propelled deep sea submersible, *Trieste*, carried U.S. Navy Lt. Don Walsh and Swiss engineer-oceanographer Jacques Piccard to a depth of nearly 7 miles (11 km) into the Mariana Trench in the western Pacific. This bulky Swiss-designed, Italian-built vessel owned by the U.S. Navy was unable to take photos or collect scientific data and could stay on the sea floor for no more than 20 minutes.

In 1964 Woods Hole Oceanographic Institution partnered with the U.S. Navy to develop a series of relatively small maneuverable deep-sea research vehicles, branded *Alvin*. Their 1976 model was certified for depths up to 13,124 feet (4,000 meters). And in 1977 *Alvin* explored the Galápagos Rift, an expedition that led to a revolution in biology. At 8,200 feet (2,500 meters) below sea level, these American geologists and geochemists found an oasis of exotic and unique animal life clustered around cracks in the ocean floor. **Hydrothermal vents** gushed shimmering, mineral-rich fluids, sometimes black, heated below by volcanic rock, earning them the nickname *smokers*.

Alvin returned to the Galápagos Rift in 1979 after being outfitted with a stronger titanium frame, this time with a team of biologists eager to explore these improbable ecosystems. *Alvin's* arms and cameras captured astonishing 8-foot-tall (2.5-meter) tube worms, bizarre crustaceans with "teeth" on their eyestalks, giant white clams with blood-red flesh, delicate flower-like relatives of jellyfish, plus unknown species of mussels, feather-duster worms, brittlestars, snails, lobsters, and more!

Imagine the excitement as this team confirmed that life, in total darkness, could manufacture food from chemicals streaming from these vents, using **chemosynthesis** instead of photosynthesis. Here, life-sustaining energy comes from the interior of the earth and not from the sun. Bacteria and other microscopic life forms at the base of this food chain oxidize hydrogen sulfide and add carbon dioxide and oxygen to produce sugar, sulfur, and water. Other creatures eat these microorganisms or offer them shelter in exchange for food they manufacture. The giant tubeworms, for example, have no eyes, mouth, or digestive system—they host symbiotic bacteria that feed them in exchange for a safe place to live.

New global studies of living bacteria in hot springs and deep sea hydrothermal vents show promising signs that life began in such inhospitable places. NASA Astrobiologist Jack Farmer has noted that biological activities of some thermophilic microorganisms leave distinctive corrosion patterns in rocks on which they grow and die. These geochemical "fingerprints" of trace fossils are subtle but detectable, which could lead to a better understanding of early life on Earth and other planets.

Recently an international team of eight earth scientists headed by evolutionary biologists at the London Centre for Nanotechnology have been studying surface rocks in Canada believed to have formed 4.3–3.8 bya near deep sea hydrothermal vents. These rocks are not only ancient, they appear to contain microfossils similar to those found in more recent vent deposits—possibly the earliest record of life on Earth. But keep in mind that claims of earliest life forms are always controversial and open to reinterpretation as new discoveries are made.

Natural Laboratories in Extreme Environments

There are more than 1,000 underwater caves in the Bahamas, the largest concentration of blue holes in the world. A network of subterranean caves connects inland blue holes with the ocean, like this one on Andros Island (**facing page**). These caves harbor poisonous clouds of hydrogen sulfide where fresh water meets salt water, a zone that nourishes sheets and floating strings of microbes—*biofilm* (**right**). This photograph was captured by the author at a depth of 100 feet near Andros Town on the island's northeast shore. Floating biofilm instantly disintegrates when touched. Scientific cave explorers see these underwater caves as natural laboratories for studying life under extreme conditions that might have existed on Earth billions of years ago.

THERMOPHILE HEAVEN

Below: Steaming runoff from Excelsior Geyser flows towards the Firehole River in Midway Geyser Basin, Yellowstone National Park, Wyoming, USA. This water averages 199ºF (93ºC), too hot for most life forms—yet "heat-loving" microbes, *thermophiles*, thrive here. Several types of colorful bacteria, including cyanobacteria, inhabit temperature-specific micro-environments in each geyser basin, and what you see in any given season represents a blend of colors, a blend of pigments. Yellow and orange pigments—*carotenoids*—protect bacteria from intense UV wavelengths in sunlight, which can be especially harsh in these flat, open places. Carotenoids trap potentially damaging wavelengths of light, energy that is utilized by chlorophyll to manufacture food. In winter, when sunlight is less intense, some of these same bacteria produce fewer carotenoids and begin to look more green than yellow.

Stromatolites

Living stromatolites offer biologists a window to the origin and evolution of life on our planet, and other planets. This famous assemblage of boulder-like living stromatolites—normally under water—is exposed at low tide in Hamelin Pool Marine Nature Reserve, Shark Bay World Heritage Area, Western Australia. Photo by Thomas B. Wilson/University of Arizona.

As of 2019, stromatolites from the 3.45-billion-year-old Strelley Pool Formation, Pilbara, Western Australia, are recognized as the oldest confirmed fossil evidence of life on Earth. Chemical studies of the Pilbara fossils using an array of high-tech imaging technologies have confirmed the presence of organic matter in these stromatolites. The striations, 2 mm apart in this specimen, are fossilized layers of minerals and biofilm, microbial mats of micro-organisms, primarily cyanobacteria. Specimen courtesy of Tom Kapitany/Crystal World, Australia.

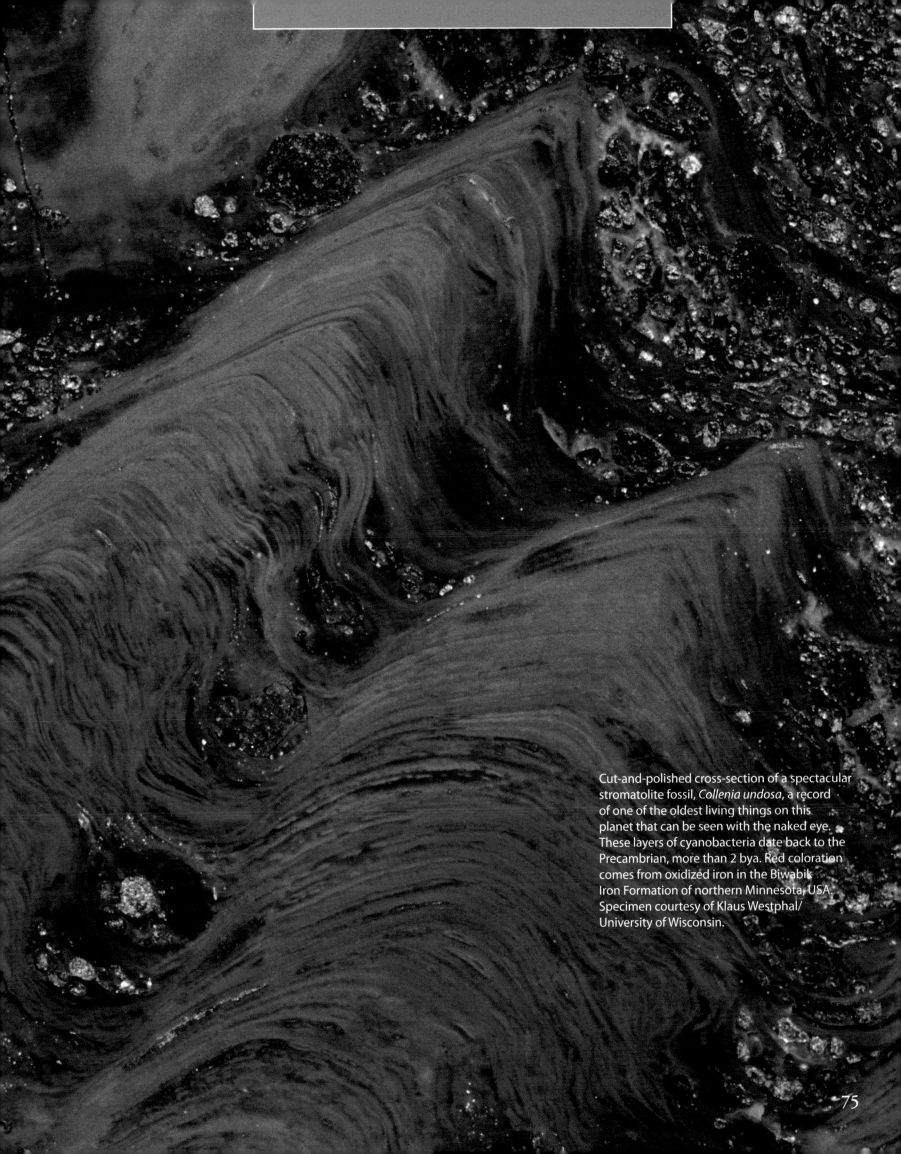

Cut-and-polished cross-section of a spectacular stromatolite fossil, *Collenia undosa*, a record of one of the oldest living things on this planet that can be seen with the naked eye. These layers of cyanobacteria date back to the Precambrian, more than 2 bya. Red coloration comes from oxidized iron in the Biwabik Iron Formation of northern Minnesota, USA. Specimen courtesy of Klaus Westphal/ University of Wisconsin.

First Multicellular Life

The leap from single-cell microbial life to multicellular Earthlings happened multiple times during the Precambrian and Cambrian. Exactly when and how these changes took place in deep time will remain a topic for research and debate for many, many years to come. But geneticists do know that this transition might have been an easy one. Studies of modern single-celled organisms have revealed that some are equipped with seemingly "unneeded" genes suited to multicellular life-styles. So although "simple," they are genetically predisposed to make the move to multicellular life, which has been demonstrated in laboratory experiments. When cells gather together in a cooperative way, specializations develop that can enhance an organism's ability to disperse, compete, and exploit environmental resources.

In Ancient Greek, **Phanerozoic Eon** means the *time of visible life*, which spans the fossil record from the Cambrian (beginning 541 million years ago) to the present day. The term was coined in 1930, long before paleontologists knew that biologically complex creatures existed earlier than the Cambrian. The convenient name **Precambrian**, which means *before the Cambrian*, covers about 4 billion years of geological history, starting with the birth of our planet's crust and ending at the onset of the Cambrian 541 mya. Our understanding of early life has been expanding with new fossil discoveries and the development of more refined dating techniques.

During the **Archean Eon** (4.0–2.5 bya) of the Precambrian, stromatolites (layered accumulations of colonial microbes and sediment) flourished in a sea of unicellular organisms. Complexity and diversification accelerated in the **Proterozoic Eon** that followed (2.5 bya–541 mya), and a full-blown "awakening" happened during the Cambrian Period (541–485 mya). Microfossils from Australia suggest that **eukaryotes** (organisms with cells containing distinct nuclei) first appeared about 1.8 bya. Multicellular life, unlike any seen today, began to dot the fossil record about 1500 mya (= 1.5 bya).

From 2 bya to 560 mya, several severe ice ages coupled with low atmospheric oxygen appear to have restrained the evolution of complex multicellular life. Virtually all life was pushed to the brink of extinction at this time, a period geologists fondly refer to as "Snowball Earth." But dramatic changes began when tectonic activity tore apart the giant continental landmass of **Rodinia** 750–633 mya, fragments of which later formed Pangea—see maps in Chapter 4. Earth's climate warmed and glaciers melted. Floodwaters fed inland lagoons and shallow seas along continental shelves, conditions favoring the expansion of cyanobacteria. These tiny photosynthesizers infused ocean and atmosphere with oxygen, setting the stage for an "explosion" of life late in the Precambrian and throughout the Cambrian.

Impressive Precambrian fossil deposits have turned up in Namibia, Australia, Newfoundland/Canada, Russia, China, and more than 25 other localities around the world. At least 50 species have been found in the Ediacaran Hills of Australia, the source of the name **Ediacaran Period** (635–541 mya), which marked the end of the Precambrian's Proterozoic Eon. Paleobiologists have described more than 200 species of multicellular Ediacaran life visible to the naked eye—names that still need to be sorted out.

Most were soft-bodied marine forms with no head, no mouth, no gut, and no eyes. Nutrient intake, waste removal, and gas exchange presumably took place directly by diffusion from and into seawater. Perhaps others were equipped with internal "fuel tanks," symbiotic single-celled organisms that could nourish their host via photosynthesis or chemosynthesis. Some Ediacarans were free-floating; the rest were confined to the sea floor. Many had leaf-like shapes, others were tubular, and some bag-like. Clear imprints of big fleshy forms as long as your arm have been found in Newfoundland rocks. And worms, some segmented, with musculature for mobility have turned up in Australia, China, and Russia. Tissues of most of these organisms are not preserved as "hard fossils"; instead, what remain are delicate traces and impressions of their soft bodies.

Given the challenges for fossilization of squishy life forms, it's no surprise that reconstructing this important transition in the history of life has been challenging. Relationships to more recent groups of animals remain obscure and hotly debated. But the deeper paleontologists dig, the more they have come to realize that Precambrian creatures were more diverse than once thought. We now know that some had body plans that were somewhat similar to well-known animals from the Cambrian. *Parvancorina*, for example, was a shield-shaped bilaterally symmetrical animal that lived on the Ediacaran seafloor, possibly a precursor to Cambrian trilobites. Others, like radially symmetrical *Eoporpita* and *Arkarua*, suggest distant kinships with corals and sea urchins. And goblet-shaped *Namacalathus* had calcified skeletal features found today only in marine animals like **brachiopods** (shelled animals that resemble clams, but are anatomically quite different).

What might have looked like a peaceful undersea garden late in the Precambrian changed dramatically during the Cambrian Period. Mobile animals with big eyes, jaws, and claws started to terrorize these placid seascapes. In response, prey animals developed defensive, calcium-rich shells and hard external skeletons. And some must have added new tools to their survival kits, like stinging cells and chemical shields—defenses not detectable in Cambrian fossils but commonly seen in their living descendants. This "arms race" is commonly referred to as the "Cambrian explosion," the first time that massive numbers of hard fossils appear in the fossil record.

The Cambrian was a period of innovation and proliferation of marine invertebrates—animals without backbones. Distant ancestors of nearly every major group of living animals on our planet made their debut before the close of the Cambrian 485 mya: brachiopods; flatworms, roundworms, and segmented worms; **cnidarians** (corals, sea anemones, jellyfish, etc.); mollusks (snails, clams, squid, chambered nautilus, etc.); **echinoderms** (sea stars, sea urchins, crinoids, etc.); **arthropods** (insects, spiders, millipedes, and crustaceans like shrimp and crabs); primitive **chordates**; and others. As embryos, all chordates have a **notochord**, a cartilaginous skeletal rod that supports the body. In vertebrate animals—most fish, amphibians, reptiles, birds, and mammals—the notochord is replaced by the vertebral column and becomes the cartilaginous discs between the vertebrae. Fish, which first appeared in the Cambrian, were eel-like and had a cartilaginous skeleton with no jaws. Plants began to move from water onto land, paving the way for animals to follow.

Although all animal **phyla** (upper divisions in the classification of organisms) were represented in the Cambrian, most of the creatures living at that time would look unfamiliar to us and

have since vanished. In 1909, Smithsonian paleontologist Charles Walcott discovered a treasure trove of "weird wonders" in the Canadian Rocky Mountains. Creatures with soft parts delicately preserved in fine-grained Burgess Shale have given humanity an unprecedented look at sea life in mid-Cambrian times 508 mya. These bizarre creatures, unlike any seen before, were given appropriately strange names, like *Hallucigenia*, named for its spiky "dream-like" appearance; *Anomalocaris*, a ferocious predator, the largest of all Cambrian animals, that grew to the length of a human leg; and *Opabinia*, a five-eyed swimmer with a snout that could probe the sea floor for food. Trilobites were also well represented at this time. Since then, similar deposits have been found in Utah, southern China, Australia, Siberia, and northern Greenland. A surprising diversity of sea anemones, jellyfish, sponges, and other invertebrates of early Cambrian age (518 mya) have turned up in China's newly discovered Qingjiang Biota.

Today, animals without backbones (= **invertebrates**) represent about 95% of all living species of multicellular creatures on Earth, and 81% of them are arthropods (90% of which are insects). Next in line are the mollusks at 8%, and coming in third at 4.5% are the chordates, the group that includes all vertebrate animals. Other invertebrate phyla each hold no more than a 2% share of the total. These percentages were clearly very different and changing through deep time, but since the Cambrian, arthropods have ruled. Given their impressive diversity and time-tested success, they will undoubtedly continue to dominate our planet well into the future.

EARLY GEOLOGIC TIME

ILLUSTRATION BY PAUL MIROCHA

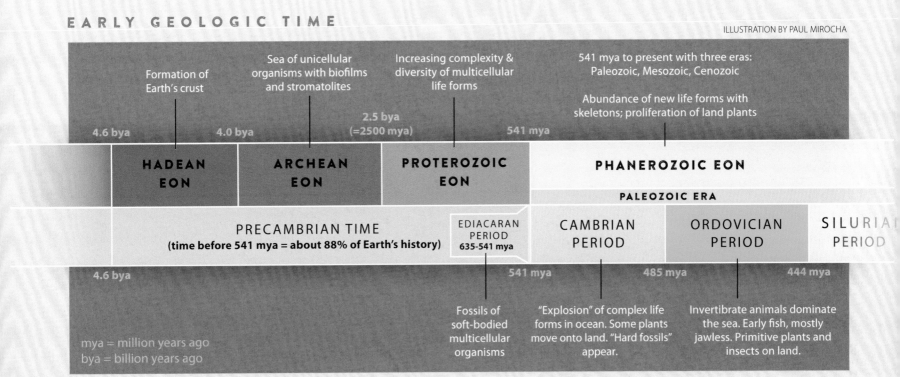

Formation of Earth's crust

Sea of unicellular organisms with biofilms and stromatolites

Increasing complexity & diversity of multicellular life forms

541 mya to present with three eras: Paleozoic, Mesozoic, Cenozoic

Abundance of new life forms with skeletons; proliferation of land plants

4.6 bya   4.0 bya   2.5 bya (=2500 mya)   541 mya

HADEAN EON   **ARCHEAN EON**   **PROTEROZOIC EON**   **PHANEROZOIC EON**

PALEOZOIC ERA

PRECAMBRIAN TIME (time before 541 mya = about 88% of Earth's history)   EDIACARAN PERIOD 635-541 mya   **CAMBRIAN PERIOD**   **ORDOVICIAN PERIOD**   SILURIAN PERIOD

4.6 bya   541 mya   485 mya   444 mya

mya = million years ago
bya = billion years ago

Fossils of soft-bodied multicellular organisms

"Explosion" of complex life forms in ocean. Some plants move onto land. "Hard fossils" appear.

Invertebrate animals dominate the sea. Early fish, mostly jawless. Primitive plants and insects on land.

Reconstruction of the sea floor as it might have appeared about 600 million years ago late in the Precambrian Era (the Ediacaran Period). This was a time of enormous biological change, a time of transition from a world dominated by microscopic organisms to one swarming with multicellular animals. Many of these early animals were either bilaterally or radially symmetrical; others were organized in spiral shapes. Fossils of these animals are unlike any known organism living or dead, but they show similarities. Some were colonial, some jellyfish-like, and some mollusk-like. Segmented forms were worm-like or trilobite-like. This diorama is based on fossils found in the Ediacara Hills of South Australia and can be seen in the Prehistoric Journey exhibit at the Denver Museum of Nature and Science, Denver, Colorado. Artwork by Chase Studio/Cedarcreek, Missouri.

EDIACARAN CREATURE

Fossilized imprint of one of Earth's first multicellular creatures, *Dickinsonia costata,* of Ediacaran age, about 560 million years old, from Nilpena, South Australia—the richest collection of early life forms in the world. *Dickinsonia* fossils are up to a meter in length. Fossilized trackways suggest that these flat segmented creatures were mobile and left imprints after absorbing nutrients from mats of bacteria. The finer the sediment grains, the clearer the impression. Photo © Jason Irving/South Australian National Parks & Wildlife Service.

TERROR OF A CAMBRIAN SEA

In its day—Cambrian times, about 505 mya—*Anomalocaris* was a top predator, a meter-long terror in what was once a peaceful place. Fossils indicate that this arthropod-like creature was a formidable hunter with big eyes and a broad band of undulating fins for the chase, fierce spiny arms for grabbing prey, and a toothy mouth for the kill. Russian artist Andrey Atuchin based this realistic interpretation on descriptions of fossil specimens from Burgess Shale deposits in British Columbia, Canada. To encounter this and other Cambrian creatures face-to-face, visit the Royal Ontario Museum's *Virtual Sea Odyssey* on the internet.

Artist's interpretation of a rare, extinct Cambrian creature, *Opabinia*, based on fossils found in Burgess Shale deposits, BC, Canada. When this animal's bizarre anatomy was fully revealed in 1972, many scientists thought it must be a hoax. Who could imagine a primitive arthropod-like swimmer with five giant compound eyes on stalks and a nozzle-like proboscis four times longer than its head! *Obapinia* was tiny—with a body about the length of your little finger. Scientists speculate that it probed the sea floor for food, and used its eyes to watch for predators. Painting by extraordinary Russian artist Andrey Atuchin.

82

BRISTLEWORMS, OLD AND NEW

Fossil bristleworm, *Burgessochaeta*, from Burgess Shale of Middle Cambrian age, British Columbia, Canada. Length of this specimen is 2.5 cm. Specimen courtesy of John Hedley/The Natural Canvas.

This marine bristleworm **below**—a bearded fireworm—is native to the tropical Atlantic ocean and the Mediterranean. It's slow-moving but well armed with spines that can inject a powerful neurotoxin if messed with. Possibly the bristly body of its ancient counterpart (**left**) was similarly equipped. Fireworm on orange cup coral photographed by Barry B. Brown in Curacao, Dutch Antilles.

Designed for Defense

With the evolution of mobile, predatory animals in Paleozoic seas came new innovations for defense—shields, spines, and stings. Sponges, corals, and *echinoderms* (crinoids, sea stars, urchins, and kin) were among the first invertebrate groups to populate the ocean. Sponges are very primitive, more like cooperative, multicellular colonies— in fact, they are so loosely organized that if ground up in a blender, its living cells will re-join to form a new sponge! Most sponges are soft-bodied filter-feeders but they are not defenseless—like corals, many are armed with toxins. Needle-like spicules also deter predators with an appetite for some sponges and soft corals. Hard corals live in a self-made fortress of calcium carbonate (limestone) extracted from seawater by their soft parts (*polyps*). Like their jellyfish and sea anemone cousins, coral tentacles are equipped with stinging cells, used for defense and capturing small prey. In contrast, many echinoderms are mobile predators, but they too have chemical defenses and tough body parts. By studying modern species, biologists have gained insight into the importance of toxins in the evolution of invertebrate animals, information that's unfortunately not preserved in Paleozoic fossils.

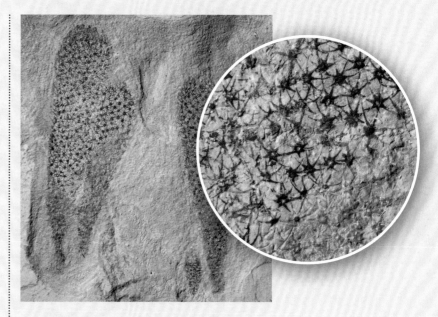

Primitive "sponge-like" animals of Cambrian age have been found in Utah. They were armed with sharp 7-point sclerites made of calcium carbonate, structural units that surely offered protection against predators. Specimen courtesy of Museum of Ancient Life/Thanksgiving Point Institute, Lehi, UT.

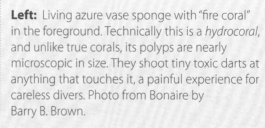

Above + Right: Much like stalked crinoids, blastoids, often called "sea buds" were echinoderms fastened to the ocean floor. They originated in the Ordovician Period, or possibly in the Cambrian, and vanished during the Permian mass extinction 252 mya. At the tip of the stem, blastoids had a hard, complex capsule (theca) with a radial array of delicate food-trapping "tentacles," seldom preserved in fossils. This life-like model and fossilized theca are on display at the Museum of Ancient Life/Thanksgiving Point Institute, Lehi, UT.

Left: Living azure vase sponge with "fire coral" in the foreground. Technically this is a *hydrocoral*, and unlike true corals, its polyps are nearly microscopic in size. They shoot tiny toxic darts at anything that touches it, a painful experience for careless divers. Photo from Bonaire by Barry B. Brown.

SOLITARY CORALS

Primitive corals were solitary, like this rugose horn coral (**above**), an extinct group of corals that were prevalent in the Ordovician Period (485–444 mya). Horn corals lived in a hard skeletal cup fastened to the sea floor at the narrow end, with tentacles circling the wide end. These two specimens of jasperized horn coral, one with a polished end, of Late Carboniferous age, were collected by Brock Sisson/Fossilogic.

To appreciate what a living horn coral must have looked like, consider each individual in this star coral colony (**left**). During the day, these fleshy tentacled polyps retreat into the safety of their stony housing. Photographed at night in Curacao by Barry B. Brown.

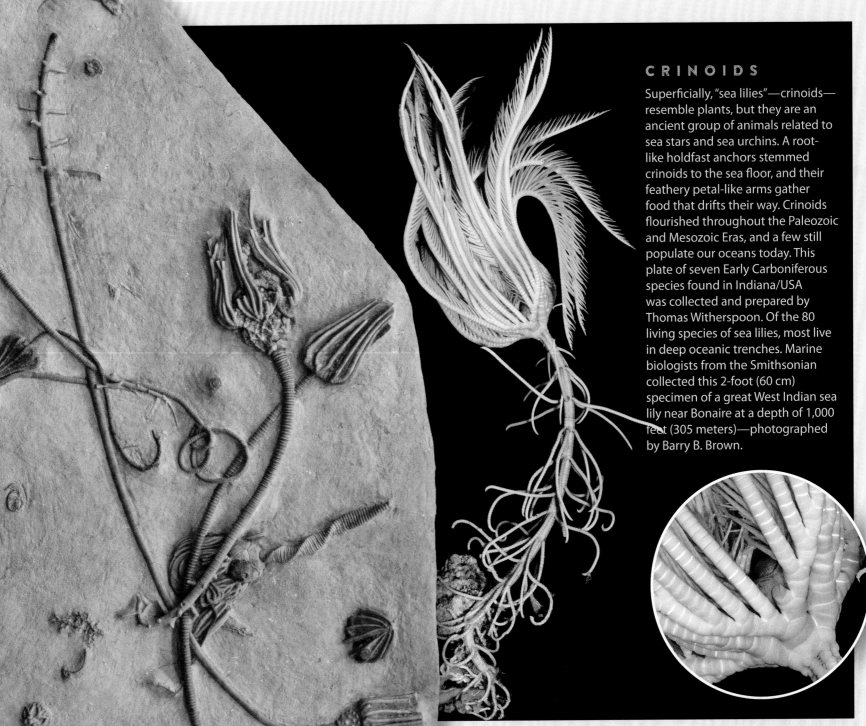

CRINOIDS

Superficially, "sea lilies"—crinoids—resemble plants, but they are an ancient group of animals related to sea stars and sea urchins. A root-like holdfast anchors stemmed crinoids to the sea floor, and their feathery petal-like arms gather food that drifts their way. Crinoids flourished throughout the Paleozoic and Mesozoic Eras, and a few still populate our oceans today. This plate of seven Early Carboniferous species found in Indiana/USA was collected and prepared by Thomas Witherspoon. Of the 80 living species of sea lilies, most live in deep oceanic trenches. Marine biologists from the Smithsonian collected this 2-foot (60 cm) specimen of a great West Indian sea lily near Bonaire at a depth of 1,000 feet (305 meters)—photographed by Barry B. Brown.

DIVING ON A SILURIAN REEF

If you could magically turn back the clock 425 million years, you could explore a shallow sea covering much of what is now the Great Lakes region of the USA. Reefs at the time were dominated by flower-like crinoids and cystoids, calcareous algae, net-like colonies of bryozoans, chain and "wasp-nest" corals, mound-shaped sponges, "clam-like" brachiopods, snails, trilobites, some early jawless fish, and predatory cephalopod mollusks, some with straight shells and others with coiled shells. This life-like reconstruction of a mid-Silurian reef offers a superb view into the past, an exhibit at the Museum of Ancient Life/Thanksgiving Point Institute, Lehi, Utah. Diorama artwork by Chase Studio, Cedarcreek, Missouri.

(1)

(2)

(3)

Trilobites—Protected & Proliferating

In the early history of complex life forms, trilobites (from *trilobos* in Greek, meaning "three lobed") are iconic. They were among the most successful sea creatures in the Paleozoic Era, known only from the Cambrian to the Permian periods (520–252 mya). Their diversity was phenomenal, with more than 20,000 described species of fossils. Some were tiny dumbbell-shaped forms without eyes or spines, smaller than a baby's fingernail. The largest, *Isotelus maximus*, grew to sizes up to 28 inches (72 cm). The middle lobe of their bodies, the thorax—is divided into segments—as few as 2 or as many as 103—most had 2–16. Many sported elaborate spines and other adornments. As in crustaceans, their hard shells offered a line of defense against predators. Most trilobite eyes are similar to those of modern bees and crabs; but one group developed a visual system unknown elsewhere in the animal kingdom. These were sophisticated compound eyes divided into separate optical units with glass-like calcite lenses and a sensory capsule behind each lens.

Morocco:
1 *Drotops megalomanicus* (Brian Eberhardie/Moussa Direct, UK)
5 *Thysanopeltis sp.* (Brian Eberhardie, UK)
7 Undescribed species, as of 2009 (Brian Eberhardie, UK)
8 *Walliserops trifurcates* (Brian Eberhardie, UK)
10 *Psychopyge elegans* (Robert Carroll/Carroll & C Enterprises, OK)
11 *Nankinolithus sp.* (Brock Sisson/Fossilogic, UT)
12 *Selenopeltis buchii* (Bill Barker/Sahara Sea Collection)

Russia:
2 *Asaphus kowalewskii* (anonymous)

USA:
3 *Peronopsis interstrictus* (Museum of Ancient Life, Lehi, UT)
4 *Huntonia oklahomae* (Robert Carroll/Carroll & C Enterprises, OK)
6 *Elrathia kingi* (Thomas Johnson/House of Phacops, OH)
9 *Homotelus bromidensis* (Geological Enterprises, Ardmore, OK)

Canada:
13 *Isotelus maximus* (Thomas Johnson/House of Phacops, OH)

(9)

(10)

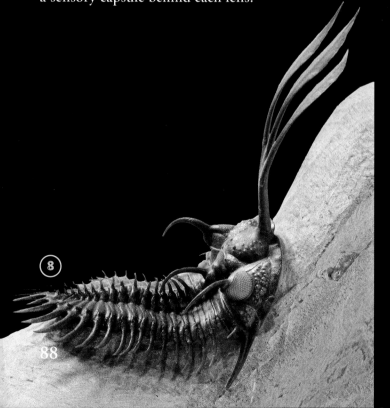

(8)

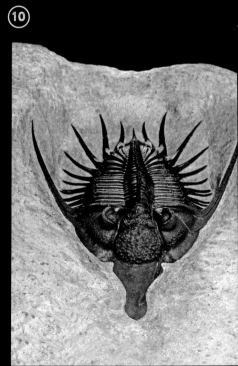

88

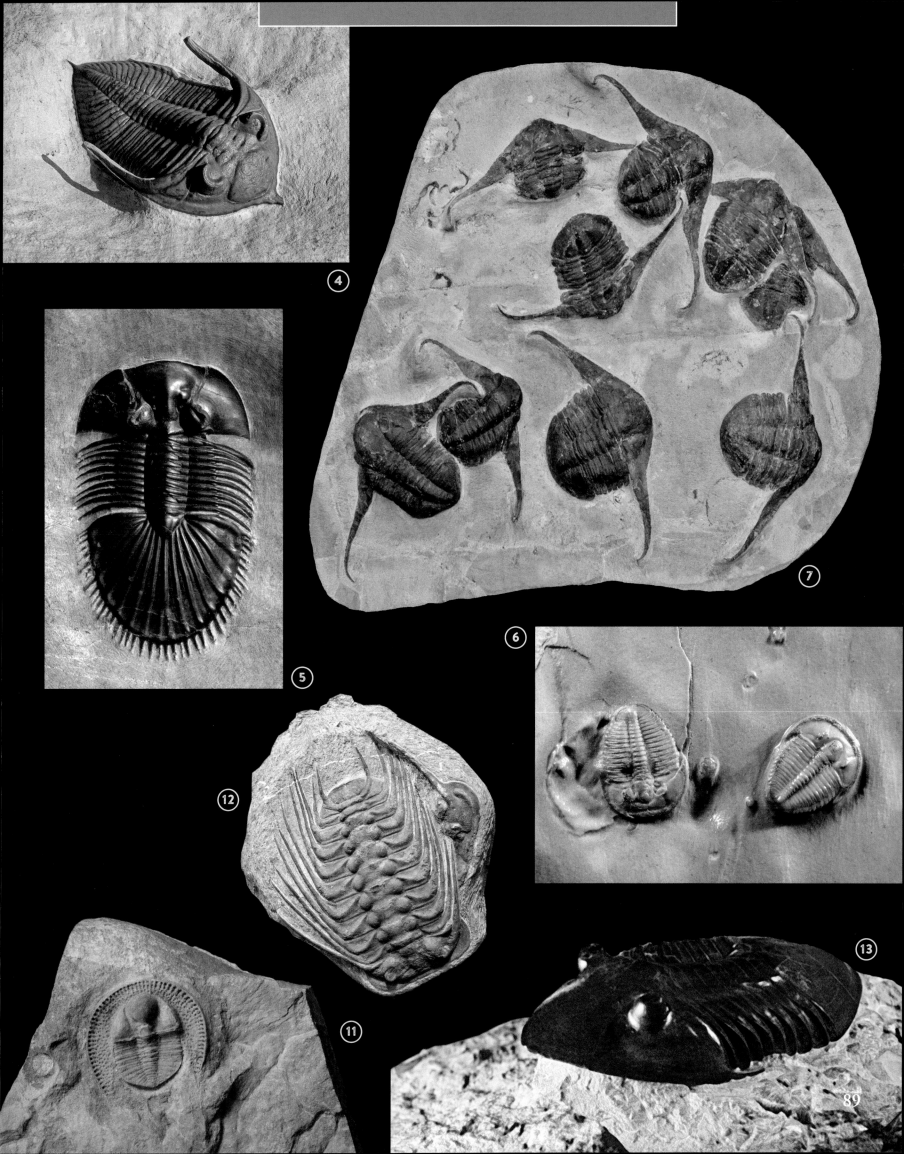

Aventuresome Arthropods

Back in Cambrian/Ordovician times, arthropods began their rise to fame. The phylum Arthropoda has been subdivided into four major groups—the trilobites; millipedes, centipedes, and insects; spiders and their relatives (notably scorpions, eurypterids/"sea-scorpions," and horseshoe "crabs"); and crustaceans (shrimp, crabs, lobsters, for example). All arthropods, past and present, have segmented bodies, jointed limbs, and a protective armor made of a tough fibrous bio-polymer, **chitin**. But paleontologists are still uncertain about many ancestral relationships among these lineages.

A CAMBRIAN CURIOSITY

Among the more mysterious early arthropods are the **Aglaspids**, known only from a sparse assortment of Cambrian-to-Ordovician fossils in North America, Europe, Morocco, China, and Australia. Although once thought to be ancestors of horseshoe "crabs," aglaspids are now recognized as more similar to trilobites. This beautiful and rare specimen—*Beckwithia typa*—of mid-Cambrian age is from the Weeks Formation in Utah, courtesy of Brock Sisson/ Fossilogic. It is 1.7 inches (4.2 cm) in total length, and during fossilization its shell was replaced with crystalline silvery-blue phosphate minerals.

Arthropods were the first marine animals to explore coastal habitats. A semi-aquatic lifestyle in the intertidal zone would have offered some arthropods a haven to feed and spawn, especially early in the Paleozoic Era before vertebrate predators gained a foothold on land. Compelling evidence from South Australia indicates that an Early Cambrian trilobite migrated into the tidal zone for synchronous breeding. And fossilized footprints of several other primitive arthropods suggest they too were crossing tidal flats more than 443 mya. Today we can witness a similar life cycle—horseshoe "crabs" gather at low tide along the shore to mate and bury their eggs in sand. During a high tide nearly a month later, the eggs hatch to begin life as larvae in shallow water. These youngsters move into deeper water as they grow.

Given their tough armor for protection and jointed appendages for locomotion, it's not surprising that marine arthropods were equipped for the move to land. This happened at various times in the Paleozoic, especially in the Late Ordovician/Silurian/ Early Devonian time frame, roughly 450–375 mya. During the Silurian Period (444–419 mya), the diversity of life in near-shore reef communities was phenomenal, setting the stage for intensified competition and predation in the sea—selective pressures that certainly helped to drive arthropods inland. Simple plants, too—like mosses and liverworts—were already established on land in the Silurian. Millipedes, scorpions, and centipedes were among the first colonizers of "dry" land, followed by spiders and insects. Primitive millipedes were almost certainly the ancestors of insects. In the grand procession of life, insects were the first and only invertebrates to master flight, which catapulted them to success. No wonder they have become the most diverse and abundant animals on our planet, by far!

A "CRAB" THAT'S AKIN TO SPIDERS

Horseshoe "crabs", better known as xiphosurans, have been around for about 510 million years. Some fossils look much like the modern Atlantic horseshoe "crab" *Limulus*—such as this one from 150 million-year-old lagoon sediments near Solnhofen, Germany (note its trackway). But as new specimens from deep time have surfaced, it has become clear that xiphosurans have had a long history of innovation, experimentation, and specialization. This can be seen in extinct species with different body plans, some of which lived in fresh water. Specimen courtesy of Raimund Albersdoerfer/Germany.

FEARSOME PREDATORS

Predatory eurypterids, often called "sea-scorpions," are well represented in the fossil record, first appearing late in the Ordovician and vanishing in the Permian. Although not true scorpions, eurypterids are relatives of spiders and scorpions. Some were small, finger-length. Others were terrors of the Paleozoic, up to 8 feet (2.5 m) —among the largest arthropods that have ever lived. Early forms were probably marine, but most fossils are found in fresh or brackish (somewhat salty) water sediments. This 5.5-inch (14-cm) specimen of Silurian age, about 425 mya, is from Herkimer County, New York, courtesy of Black Hills Institute of Geological Research.

TRUE SCORPIONS

So far, the oldest known scorpion is about 440 million years old, found in Wisconsin west of Milwaukee, an area that was covered by a warm shallow sea during the Silurian Period. This ancient species had more belly plates than today's scorpions, but its well-preserved internal anatomy is quite similar. By considering modern scorpions, we can speculate—without fossil evidence at this point—that scorpions were the first land animals to care for their newborns. This Arizona bark scorpion gave birth to 11 fat infants that promptly crawled onto her back— some hidden in the pile. A week later, after their first molt, the young "shape-shifted" into miniature adults and left mom to fend for themselves.

Scorpion fossil from Brazil's Crato Formation, a 108-million-year-old limestone deposit of early Cretaceous age that contains some of the world's most diverse and best-preserved arthropods. Specimen courtesy of Merv Feick/Indiana9 Fossils.

LEGS ON LAND

Millipedes and centipedes are closely related leggy creatures that were among the pioneers to explore land out of water. Both lack the waxy protective layer in their body armor that most other arthropods possess, so they must avoid the sun and dry places. Even giant desert centipedes, like this 8-inch (20-cm) *Scolopendra* (**below**), are strictly nocturnal and spend most of their time in humid burrows or crevices. Centipedes are fast-moving hunters equipped with two legs per segment and a pair of large venomous pincers. In contrast, millipedes (**top**) have four legs per segment and are never in a hurry. They prefer to eat decaying organic matter. When conditions are unfavorable for feeding or reproduction, desert millipedes—like this Arizona species—curl up in an underground retreat and go dormant, for many months if necessary.

CRUSTACEANS

Crustaceans are the most abundant arthropods in virtually every aquatic environment and are a vital part of ecological food webs—including our fishing industry. Major crustacean groups all appeared in the fossil record during the Paleozoic Era, but many relationships remain unclear. At some point in history, at least one lineage of crabs traded gills for lungs and life on land—exemplified by today's Caribbean land crab *Gecarcinus ruricola* (shown below). Nevertheless, they must still migrate to the ocean for spawning.

Over the past 380 million years, fairy shrimp have mastered the art of survival in hostile environments. Some can tolerate temperatures as high as 104°F (40°C) and water three times saltier than seawater. Sonoran Desert fairy shrimp—shown below with a brood pouch full of eggs—can miraculously appear a day or two after a monsoon storm. Their eggs can survive for years in dry mud, and upon hatching, these mini-shrimp must grow and reproduce before the pond dries up—often a race against time.

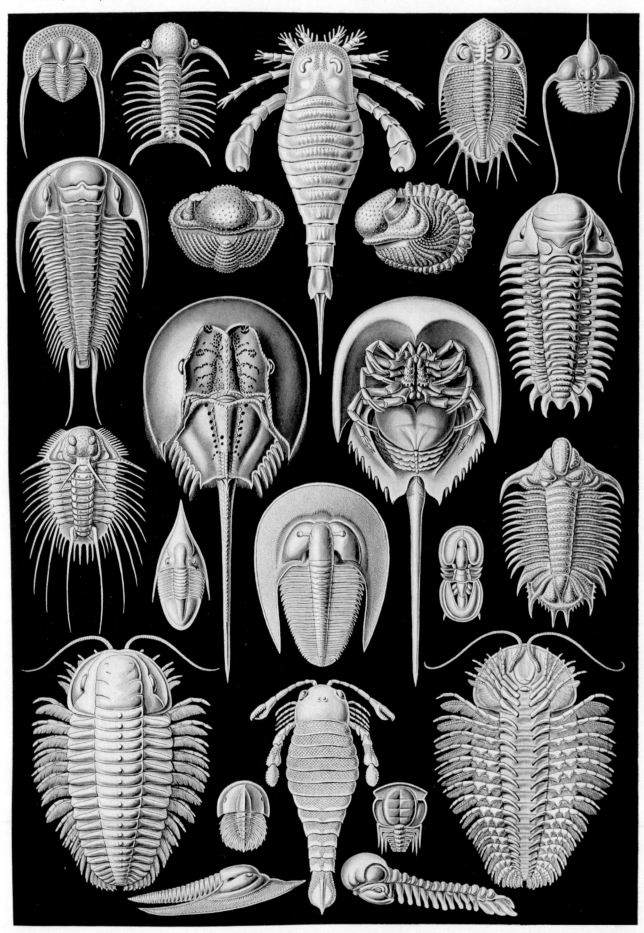

Aspidonia. — Schildtiere.

TRILOBITES, SEA "SCORPIONS" (center, top & bottom),
& HORSESHOE CRABS (center, right & left).

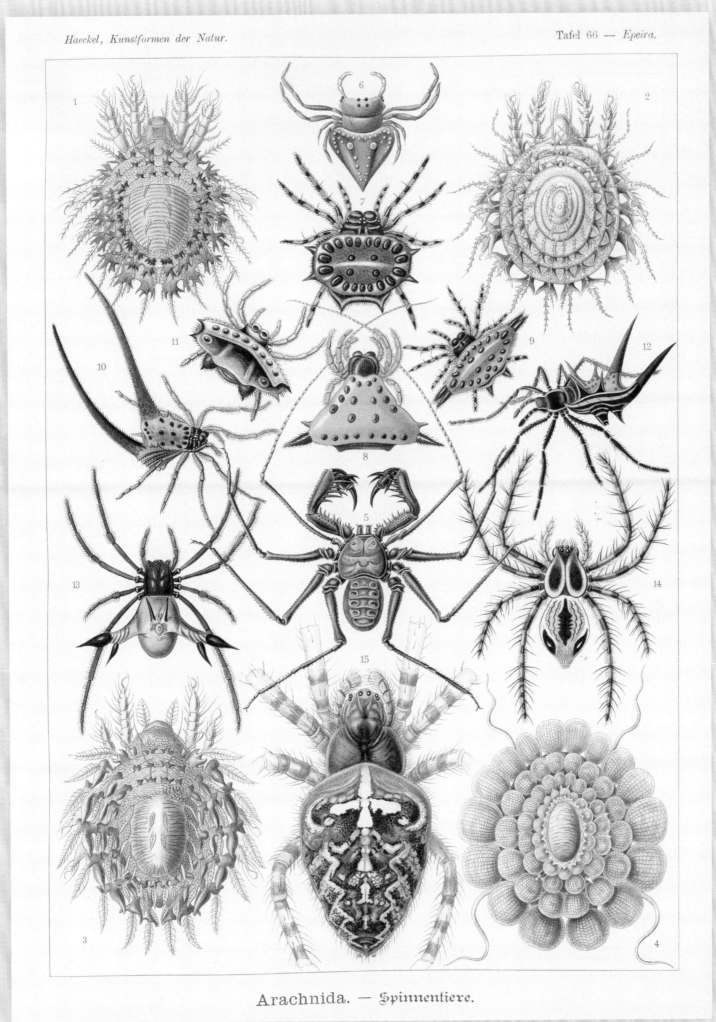

Arachnida. — Spinnentiere.

Squids, octopuses, and cuttlefish have no protective housing but these agile swimmers can instantly change colors. And if seriously threatened, they can release a dense cloud of black ink to aid their escape. A complex array of nerves and muscles tell pigment cells—*chromatophores*—to expand or contract for camouflage, defense, or courtship. And reflecting cells—*iridophores*—create iridescent greens, blues, silvers, and golds. Some squid use red to attract females and white to repel other males. They can even split their body coloration—the red half facing a female, while keeping the other half white to fend off males! Barry B. Brown photographed this Caribbean reef squid in Curacao.

Sea Shells—Old, New, Lost

Mollusks (Phylum Mollusca) are among the most familiar animals on Earth, second only to the arthropods in abundance and diversity. Their history is also rooted in the Paleozoic Era (541–252 mya). Biologists recognize three major classes of mollusks: Bivalvia—clams, mussels, oysters, and scallops; Gastropoda—snails and slugs; and Cephalopoda—squids, octopuses, cuttlefish, nautiloids, and extinct ammonoids. Although hugely variable in size, shape, and texture, all mollusks have a ribbon-like rasping tongue—a *radula*—for food gathering, and a mantle, a layer of shell-producing tissue with a cavity below it that holds the gills and other key organs. Some mollusks—squids, for example—have no external shell, only a thin flexible internal rod-like *pen* for support and muscle attachment. Most slugs also have remnants of a shell, but some have none at all.

Many early naturalists believed that brachiopods (= lamp shells) should be grouped with mollusks because they resemble clams. But after taking a closer look at extinct and living species, biologists have changed this view. Brachiopods have a unique anatomy and life style, so they are now classified in a separate phylum—Phylum Brachiopoda. Unlike clams and most other bivalve mollusks, brachiopods have two shells of unequal size, held together by muscles or hinged with paired teeth and sockets. Clams have twin shells of equal size with a different hinge design. Brachiopods have also developed a peculiar feeding apparatus—a lophophore. This hollow, rounded or spiral-shaped organ with tentacles is covered by hair-like cilia that bring seawater with food particles to the mouth. And except as larvae, brachiopods are sedentary. A few live in burrows, but most live permanently attached by a fleshy stalk to something hard on or above the sea floor. In contrast, as anyone who has dug for clams knows, these mollusks prefer to be buried, and some are surprisingly mobile. Clams can actively dig with their muscular foot, and scallops—their close relatives—can propel themselves to safety with rapid shell-clapping movements.

Fossilized carpet-shell clams, bivalve mollusks—mostly *Tapes decussatus*—of late Pliocene age (about 3 mya) from Piacenza, Italy. Clam shells have two equal halves (= *valves*). This species is still alive and a prized food item in Italy. Specimen courtesy of Martin Goerlich/Eurofossils.

Snails are typical gastropod mollusks, at home in nearly all aquatic and terrestrial environments. Marine species create the fancy coiled shells that beachcombers love to collect. Every baby snail that hatches from an egg has a soft shell attached to its twisted body. The snail must consume enough calcium to strengthen and enlarge its shell. Like turtles, a snail can never crawl out of its shell, and the shell grows with the animal. But unlike a turtle's shell, the snail's shell is a non-living coat of armor. As the snail grows, it adds layer upon layer of calcium carbonate minerals around the lip of the opening, like circular building blocks. All snails can retreat within their shells, sealed by a tough door-like *operculum* for protection against predators and environmental extremes. Under persistent drought, some land snails can remain encased in their shells and lapse into a state of dormancy for several years.

Early in the Paleozoic Era, all major groups of mollusks were represented in the fossil record, but their shells changed dramatically over time. In an ambitious EU-funded study of the "micro-structure" of mollusk shells, a research team discovered that Early – Middle Cambrian (541–509 mya) shells contained more organic matter and were relatively thinner and more flexible than today's shells. Back then, few predators were armed with crushing claws and jaws. But when mobile predators that could break, drill, and pry open shells proliferated in the Ordovician, about 450 mya, mollusks responded. They evolved thicker, more mineralized shells strengthened with a tough layer of *nacre* (commonly called "mother of pearl"). And by the Carboniferous (359 mya), most mollusk lineages were already armed with shells similar to those seen today.

BRACHIOPODS

Oriental lamp shell, *Lingula anatina*, from the Ariake Sea, Fukuoka, Japan—one of about 300 living species of brachiopods. *Lingula* is an ancient group that dates back to the Cambrian. Its thin shells have no hinge—muscles hold the two parts together. This burrow-dweller attaches itself to the sea floor by a fleshy stalk (= *peduncle*). Specimen courtesy of Leon Theisen/Custom Paleo.

Fossil lamp shell, *Cyrtospirifer rudkinensis*, one that clearly shows a typical brachiopod shell structure, its two "halves" being unequal in size. The name "lamp shell" comes from the resemblance of some species to Roman oil lamps. *Cyrtospirifer* are only known from Middle and Upper Devonian deposits. This shell was found in Voronez, Russia. It is nearly 2 inches (4.7 cm) wide.

If there's a pinnacle in the evolution of marine invertebrates, cephalopods must surely be at the top. Cephalopods are formidable predators equipped with eight or more arms, a beak-like radula, and jet propulsion, achieved by forcefully squirting water from the mantle cavity. And their nervous systems are exceptionally complex, although less centralized than in vertebrate animals. Of roughly 500 million neurons in an octopus, for example, only 150 million are in its brain. The remaining 350 million radiate out in clusters of *ganglia* along each of its eight arms. By comparison, rats have a total of about 200 million neurons and most mollusks only 20,000. Much of this neural circuitry in cephalopods is devoted to vision and camouflage. Their large eyes match the complexity of the human eye, and the arms of an octopus can independently touch, taste, and control movements without input from the brain. No wonder they are so adept at solving puzzles and changing skin color and texture in a flash!

TWISTED BODY DESIGN

This Cuban painted snail, *Polymita picta*, is pooping from its head end. In fact, the animal's anus opens near its mouth and breathing hole! This twisted anatomy results from a developmental process called *torsion*. In land snails, larval growth takes place within the egg; in aquatic snails, the larvae are free-swimming. At first, these larvae are bilaterally symmetrical. But during development the inner organs and tiny shell rotate and end up in unusual positions relative to the foot. All gastropods mollusks do this, but some— slugs for example—undergo a second developmental stage called "detortion." The organs in their larval bodies return to a linear, bilaterally symmetrical orientation. As a result, these "snails without shells" poop as expected, from the rear end.

SEA SLUG PARTNERSHIP

Some of the most colorful animals in the ocean are slugs, snails without much of a shell if any. This lettuce sea slug, *Elysia crispata*, eats green algae, but not all of it is digested—some of it is given a home in the slug's intricately folded lettuce-like appendages that cover its back. In this intimate symbiotic relationship, the plant cells manufacture food that helps to energize this sea slug. Photographed by Barry B. Brown in the Caribbean at Bonaire Island.

Fossils suggest that early cephalopod ancestors had chambered shells. These shells were specialized for buoyancy control by allowing the mollusk to adjust fluid and gas levels within the chambers. The earliest shells, dating back to the Late Cambrian Period (515 mya), were long, straight, and pointed like ice cream cones—they belonged to distant relatives of today's chambered *Nautilus*. In the Late Silurian-Devonian, about 400 mya, another major lineage arose, the ammonites—they, too, probably descended from primitive straight-shelled nautiloids. Many sported curved and loosely coiled shells. Others had tight coils fused into a circular disk. Studies of shell geometry suggest that cephalopods with straight, hook-like, or loosely coiled shells were relatively slow-moving and occupied quiet marine environments. In contrast, disk-shaped shells—typical of most ammonites during the Mesozoic Era (252–66 mya)—offered better buoyancy control for mobility. Ammonites perished along with dinosaurian giants at the end of the Mesozoic. In spite of a rich fossil record, much disagreement about relationships has persisted. But we've come a long ways since Medieval times when people believed that ammonite fossils were "snakes turned to stone"!

HETEROMORPHIC CLUSTER

Natural death assemblage of loosely coiled spiny nautiloids and ammonites (several species) of Cretaceous age from Morocco. Commonly called *heteromorphs* because their conical shells are twisted into a wide variety of shapes, and none are tightly coiled. Specimen courtesy of Burkhard Pohl/Granada Gallery, Tucson, Arizona.

STRAIGHT-SHELLED NAUTILOIDS

Primitive nautiloids with long, straight, conical shells–– mid-Paleozoic *orthocone* cephalopods—are abundant in Morocco. This is a typical preparation, with the black limestone slab cut and polished to expose portions of the chambered shells. Specimen courtesy of Bill Barker/Sahara Sea Collection.

Above: Model of a coiled ammonite in a Mesozoic sea. Diorama at the Museum of Ancient Life/Thanksgiving Point Institute, Lehi, Utah. Artwork by Chase Studio.

Left: An ammonite fossil that shows the animal's thin, outer, mother-of-pearl shell layer flaking away to reveal two underlying layers. On the innermost layer, you can see intricate, reticulated suture patterns, which are much like a "fingerprint" for each species. This Late Cretaceous ammonite, *Placenticeras meeki*, is from South Dakota. Specimen courtesy of the Black Hills Institute of Geological Research.

Like snails, nautiloid/ammonite shells were permanently attached to the animal's soft squid-like body, which occupied the wide, end-chamber of its shell. During growth, the animal added newer, larger chambers to the coil. Neighboring flotation chambers were separated by intricately folded walls (septae), and where fused to the outer shell, we see distinctive "joint patterns." These suture patterns are useful in ammonite classification because they vary from species to species. Specimen courtesy of Lars Berwald.

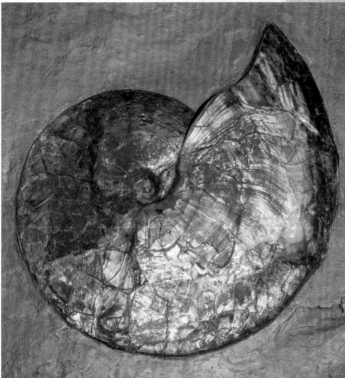

SHELL OPALESCENCE/ IRIDESCENCE

Below: Opalescent (= milky iridescence) ammonite fossil, *Cleoniceras besairiei*, of Early Cretaceous age, from Mahajanga Province, Madagascar. Diameter of this ammonite specimen is 3.5 inches (9 cm). It has been mounted on a rock containing an iridescent layer of *Placenticeras* ammonite shell from Meade County, South Dakota. Specimen courtesy of Neal Larson/ Larson Paleontology Unlimited.

AMMOLITE

Rare and beautiful fossils of the ammonite *Placenticeras intercalare* are highly prized by collectors. Geological processes transformed the mother-of-pearl (aragonite) of the animal's the inner shell into a colorful iridescent material officially recognized as a precious gemstone, *ammolite*. This species lived during the Late Cretaceous (110–66 mya); collected from Bearpaw Shales of Alberta, Canada. The diameter of this specimen is 20 inches (51 cm). Specimen courtesy of Canada Fossils, Ltd.

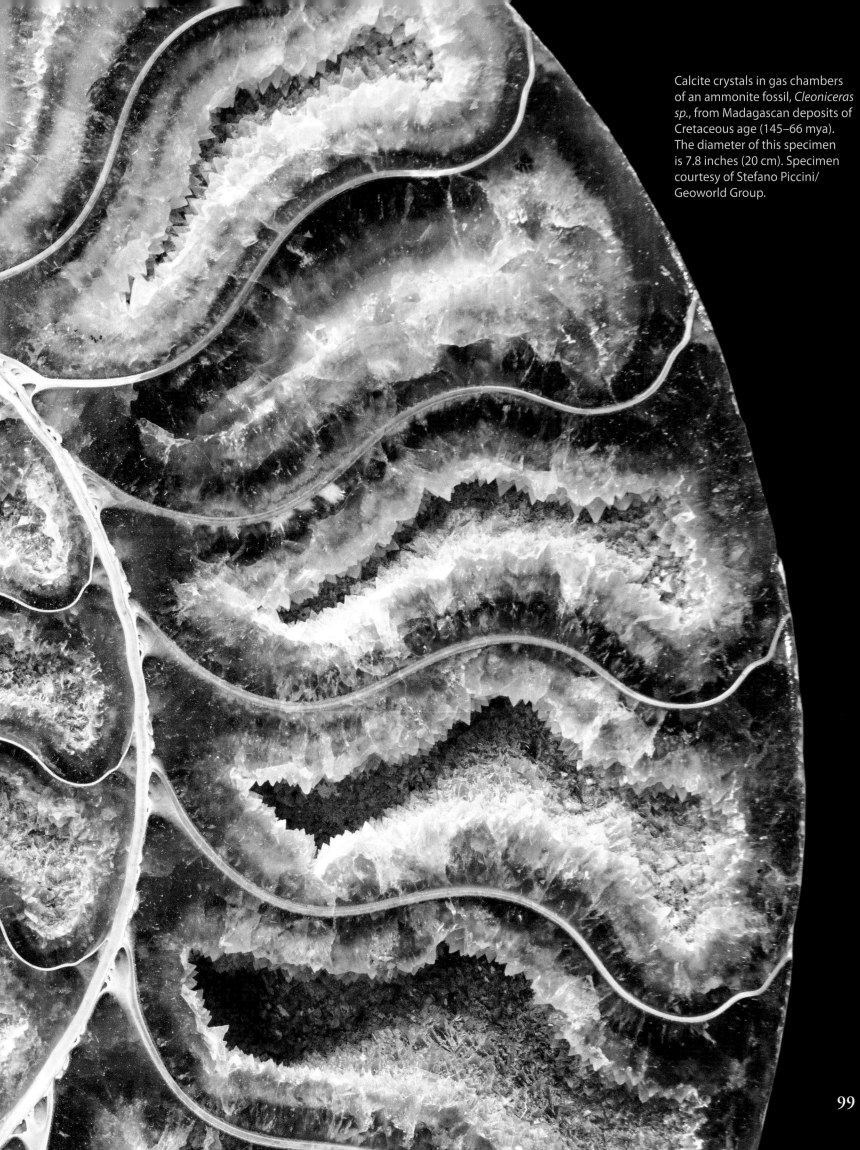

Calcite crystals in gas chambers of an ammonite fossil, *Cleoniceras sp.*, from Madagascan deposits of Cretaceous age (145–66 mya). The diameter of this specimen is 7.8 inches (20 cm). Specimen courtesy of Stefano Piccini/ Geoworld Group.

Primitive animals with backbones first appeared in Cambrian seas. Their basic body plan was similar to that of all modern vertebrates—they had a notochord, rudimentary vertebrae with muscles, and a well-defined head and tail. The earliest "fishy forms" had a cartilaginous skeleton, gills, and a mouth with rasping teeth, but no jaws, no stomach, and no paired fins. In most respects they resemble modern lampreys and hagfish, which are soft-bodied jawless fish without scales, the **Agnatha**. Hagfish are unusual and even more primitive than lampreys. They can feed through the mouth and process food within their tube-like digestive system or absorb dissolved nutrients directly through the skin like their aquatic invertebrate ancestors.

Other major groups of jawless fish, now extinct, were well armored. They swam in both marine and freshwater environments 500–360 mya, beginning in the Cambrian and ending in the Devonian. Some had a bony head shield and others had bony plates in their skin (dermal scales) or bodies covered with small spiny scales. One group had paired pectoral fins, well-developed dorsal and anal fins, and a broad flat tail fin shaped for active swimming—features seen in most modern fish.

And then came spiny shark-like fish, the **acanthodians**, about 440 mya—equipped with jaws and fins reinforced with prominent bony spines. These primitive "stem sharks" are extinct members of a diverse and poorly understood group of fish, with a blend of features common to cartilaginous fish (sharks and rays) and their bony cousins. They were small with internal skeletons made of cartilage. But they also had a bony flap over the gills, fins with a bony base, and diamond-shaped scales—resembling the scales on modern gar. Some had no teeth, while others had teeth fused to the jaw. Aside from the 80+ species of living lampreys and hagfish, jaws became a permanent fixture in the procession of all vertebrate life that followed.

During the Devonian, jawed fish underwent dramatic diversification into many shapes and sizes. Those with heads covered with thick bony plates, the **placoderms** (meaning *plate-skinned*), rose to dominate and terrorize the seas. Some, notably *Dunkleosteus*, grew to lengths matching that of today's largest great white sharks, but instead of having teeth, their self-sharpening, plated jaws served as blade-like slicers. Some of the smaller armored fish were bottom dwellers with tooth plates presumably used to crush mollusks, crustaceans, and other invertebrate animals. The dramatic rise of placoderms during the Devonian came to an abrupt end after about 50 million years, having reached an "evolutionary dead end" brought on by a mass extinction event 359 mya.

Two other major groups of fish survived the Devonian/Carboniferous mass extinction and have continued on a path to success for more than 400 million years: cartilaginous fish and bony fish. Sharks, skates, and rays all have skeletons made of cartilage and they have no swim bladder (an internal gas-filled sac that allows them to regulate their buoyancy)—if sharks don't keep moving, they will sink to the sea floor. In the absence of placoderm fish, shark diversity blossomed and they mastered the sea during the Carboniferous (359–299 mya). Most of these weird and wonderful species are now extinct, but others have taken their place. The largest of all sharks, **megalodon**, resembled a great white on steroids—but the two species were not close relatives. Megalodon ruled for about 13 million years, and paleontologists have determined that its downfall coincided with rising numbers of great white sharks about 3–4 mya. Did these smaller, more agile sharks outcompete their bigger brethren? Possibly.

Bony fish have been even more successful than cartilaginous types. They have at least some bone in their skeleton and most have a swim bladder. This group includes **lobe-finned fish** (**crossopterygians**), with fleshy limb-like fins, and **ray-finned fish**, those with fins that contain branching bony rays that stem from the base of each fin. We all know ray-finned fish that we catch for sport or serve for dinner. They have continued to balloon in numbers over the last 250 million years and are now the largest group of fishes, 27,000+ species strong—roughly half of all living vertebrates!

Scientists have concluded that a common ancestor to ray-finned and lobe-finned fishes could breathe underwater with gills or at the surface with lungs in low oxygen environments. In the ray-fin lineage, lungs evolved into the swim bladder. "Lobefins" on the other hand, kept both lungs and gills, as seen in modern lungfish. And since the discovery of a living **coelacanth** in 1938 in deep water off the coast of South Africa, along with an Indonesian species found in 1997, biologists have been able to probe deeper into their past. These fish have gills and remnants of lungs—lungs that would no longer be of use in the oxygen-rich ocean depths where these fish have survived.

Lungfish are the closest living relatives of four-legged vertebrates (**tetrapods**), so the fin anatomy of their fossilized, Paleozoic ancestors is of special interest. Their paired, muscular pectoral and pelvic fins attach to the body with a single bone—the *humerus* in the forelimb, the *femur* in the hind limb—an anatomical development that has proven beneficial for life on land. In water, gravity is less of a problem; but on land, a skeleton must be flexible and able to support the animal's weight while walking. Numerous studies of extinct and living tetrapods have allowed evolutionary biologists to trace the progression of these changes. There is no longer any doubt that specialized fins evolved into legs of the first land vertebrates—the amphibians—and the limbs of all tetrapods to follow, including *Homo sapiens*.

100

JAWLESS FISH

Models of both fish created by artist Charles McGrady; on display at the Wyoming Dinosaur Center, Thermopolis, Wyoming, USA.

Above: Model of a primitive, eel-like jawless fish without paired fins, *Jamoytius kerwoodi*, which lived about 444 mya in the Silurian Period. Most were only about 7 inches (18 cm) long, although some grew to twice that size. This early fish with weakly mineralized scales had well-developed eyes, one nostril, 10 or more gill slits, and a circular mouth without teeth—it was probably a filter-feeder or scavenger.

Above: Model of a *Drepanaspis gemuendenensis*, a heavily armored jawless fish that swam in Early Devonian seas at the onset of the Age of Fishes (419–359 mya). This fish belonged to an extinct group of about 300 species—**heterostracans**—that lived in sandy lagoons in North America, Europe, and Siberia—probably bottom-feeders, like modern rays. These fish had bony plates in their skin, one pair of gill openings, no calcified internal skeleton, and no paired fins. Most were small, typically less than 12 inches (30 cm) in length.

ARMORED FISH WITH JAWS

Skull of a giant Devonian placoderm fish, *Dunkleosteus marsaisi* (**above**), being prepared by a technician in the Bern Natural History Museum, Switzerland. This terror of Late Devonian seas was a top predator with sharp bladed jaws and no teeth. It had an enormous head with hinged body shields, skin without scales, and an internal skeleton of mostly cartilage with a bony pelvis. Biomechanical reconstructions of *Dunkleosteus* head bones and muscles suggest that these fish had the most powerful bite of all fishes, extinct or living. Specimen from Maider, Morocco. Photo © Lisa Schäublin, courtesy of the Bern Natural History Museum. The model, created by Serge Xerri of Rabat, Morocco, shows the fish's front half without skin.

Prehistoric placoderm fish, *Coccosteus cuspidatus*, in Old Red Sandstone of Mid-Devonian age (393–383 mya), from Caithness, Scotland. This small species, about 12 inches (30 cm) long, preferred fresh water. Specimen courtesy of Chris Moore Fossils.

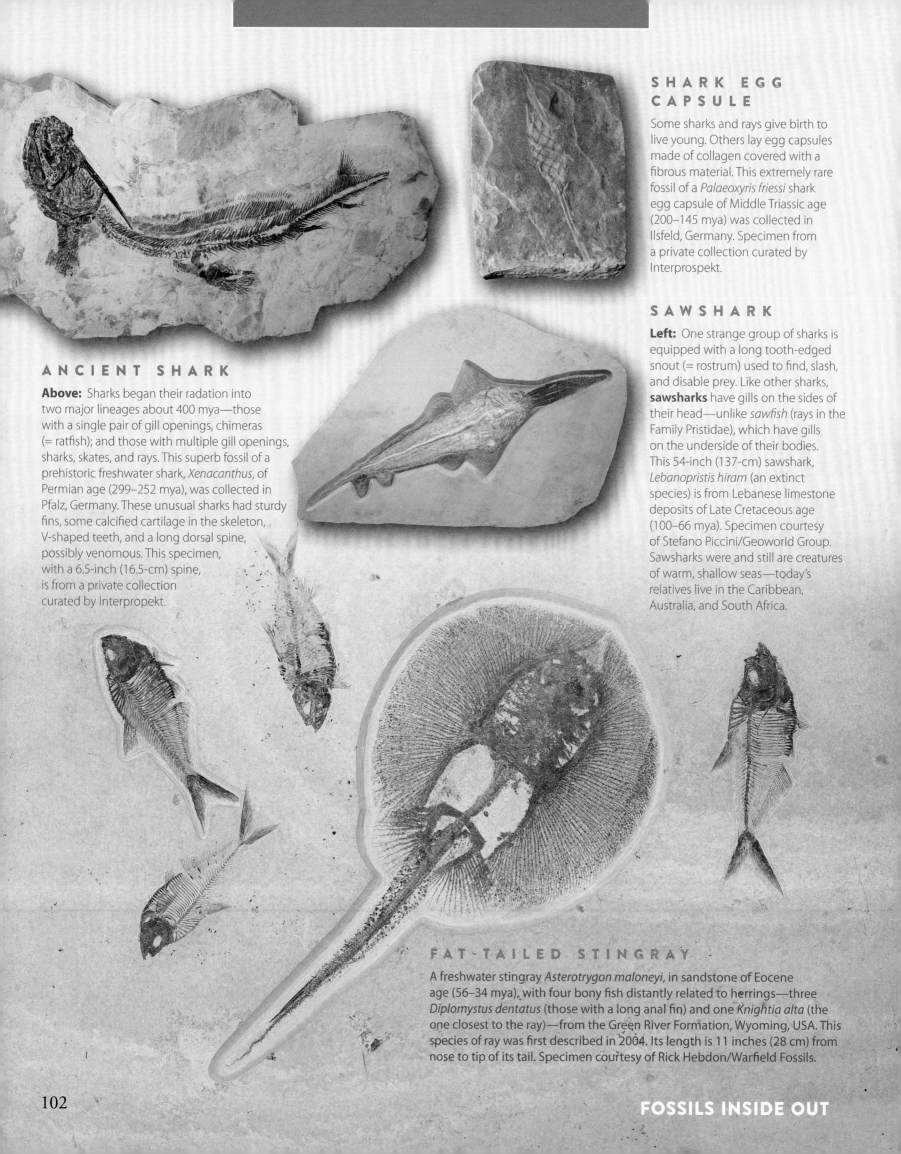

SHARK EGG CAPSULE

Some sharks and rays give birth to live young. Others lay egg capsules made of collagen covered with a fibrous material. This extremely rare fossil of a *Palaeoxyris friessi* shark egg capsule of Middle Triassic age (200–145 mya) was collected in Ilsfeld, Germany. Specimen from a private collection curated by Interprospekt.

SAWSHARK

Left: One strange group of sharks is equipped with a long tooth-edged snout (= rostrum) used to find, slash, and disable prey. Like other sharks, **sawsharks** have gills on the sides of their head—unlike *sawfish* (rays in the Family Pristidae), which have gills on the underside of their bodies. This 54-inch (137-cm) sawshark, *Lebanopristis hiram* (an extinct species) is from Lebanese limestone deposits of Late Cretaceous age (100–66 mya). Specimen courtesy of Stefano Piccini/Geoworld Group. Sawsharks were and still are creatures of warm, shallow seas—today's relatives live in the Caribbean, Australia, and South Africa.

ANCIENT SHARK

Above: Sharks began their radiation into two major lineages about 400 mya—those with a single pair of gill openings, chimeras (= ratfish); and those with multiple gill openings, sharks, skates, and rays. This superb fossil of a prehistoric freshwater shark, *Xenacanthus*, of Permian age (299–252 mya), was collected in Pfalz, Germany. These unusual sharks had sturdy fins, some calcified cartilage in the skeleton, V-shaped teeth, and a long dorsal spine, possibly venomous. This specimen, with a 6.5-inch (16.5-cm) spine, is from a private collection curated by Interpropekt.

FAT-TAILED STINGRAY

A freshwater stingray *Asterotrygon maloneyi*, in sandstone of Eocene age (56–34 mya), with four bony fish distantly related to herrings—three *Diplomystus dentatus* (those with a long anal fin) and one *Knightia alta* (the one closest to the ray)—from the Green River Formation, Wyoming, USA. This species of ray was first described in 2004. Its length is 11 inches (28 cm) from nose to tip of its tail. Specimen courtesy of Rick Hebdon/Warfield Fossils.

MOONFISH

An elegant moonfish, *Mene rhombea*, a marine species that lived during the Middle Eocene (48–40 mya) of the Cenozoic Era. It belongs to the largest and most diverse group of bony fishes, the Perciformes. This fossil was found at Monte Bolca, Italy. Specimen courtesy of Mineralien Zentrum Andreas Guhr.

Viper fish, *Eurypholis boissieri*, a 95-million-year-old bony fish from Hjoula, Lebanon. This small predator is easily recognized by the 3–4 prominent bony plates on its back. The larger of the two fish is 7.25 inches (18.5 cm). Specimen courtesy of Stefano Piccini/Geoworld Group

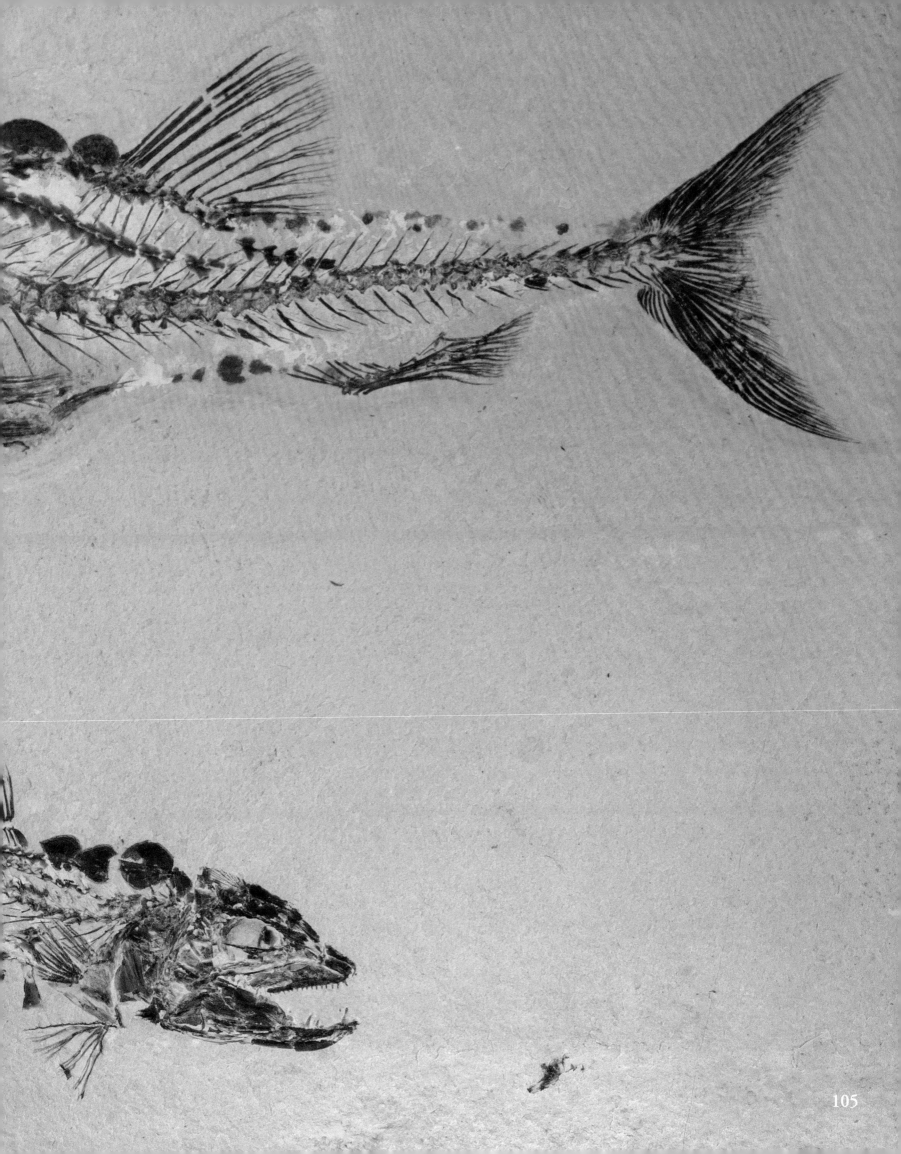

AMPHIBIANS ON THE MOVE

Early fish with limbs and lungs had a great advantage in being able to escape aquatic predators, stagnant pools, or drying ponds. A few modern species—"walking" catfish and mudskippers, for example—can move considerable distances overland by wiggling and using their muscular pectoral fins to push themselves along as if on crutches. They can breathe through their skin and have specialized chambers for air storage. Lungfish even have primitive lungs that function much like those of amphibians. Although these ancient fish bear little resemblance to modern toads, frogs, salamanders, and their worm-like cousins the caecelians, it's clear that adaptations for movement on land have evolved multiple times.

Amphibians were well represented in the fossil record late in the Carboniferous Period, 315–299 mya. By this time a few had reached monstrous sizes, some the length of today's alligators—they even had thick, scaly or plated skin, a barrier against dehydration. Others, like *Eryops*, were built like "sci-fi war machines" with stout 200-pound (90-kg) bodies, short strong legs, and massive jaws with blade-like teeth. In the absence of competition, this worldwide group of huge predatory amphibians, the **temnospondyls**, flourished in Earth's swamplands for 210 million years.

Although amphibians were moving away from a purely aquatic lifestyle, most retained a physical dependence on wet places. For reproduction they needed to be near or in water. Amphibian eggs, similar to fish eggs, have no shell—the embryo is protected within a soft, jelly-like capsule. Upon hatching, the gilled, aquatic larvae acquire lungs and legs as they grow, allowing some to metamorphose into landlubbers.

Creatures better adapted to life on dry land and a changing physical environment pushed many of these strange salamander-like giants to the edge of extinction. Early reptiles—some with mammal-like skulls—entered the scene late in the Carboniferous, about 315 mya, and expanded rapidly on land during the Permian (299–252 mya). These new arrivals packed a revolutionary weapon in their arsenal of survival tools: the **amniotic egg**. Protected by leathery or mineralized shells, amniotic eggs could be laid far from water, an evolutionary game-changer (see the next section, *The Miraculous Egg*). For temnospondyls, this reptilian invasion was pretty much the end of life on land. Amphibians that persisted had retreated to rivers, lakes, and lagoons.

Few amphibians survived the mass extinction event at the end of the Permian. Studies of their skeletal anatomy suggest that those that made it into the Triassic (252–201 mya) were fully aquatic as adults. But by mid-Triassic times, water-worlds had been invaded by an array of terrifying swimming reptiles—huge crocodile-like phytosaurs, long-necked plesiosaurs, dolphin-like ichthyosaurs,

FISH OUT OF WATER

Below: Model of *Tiktaalik*, an extinct sarcopterygian (lobe-finned) fish from the Late Devonian, about 375 million years ago. These fish share many anatomical features with amphibians and other tetrapods (four-legged animals) and are among the first to have ventured onto land. Well-preserved fossils were discovered in the Canadian Arctic. *Tiktaalik* was over 3 feet (a meter) in length and some fossils suggest it was more than twice this size. This model, create by artist Charles McGrady, is displayed at Wyoming Dinosaur Center, Thermopolis, Wyoming, USA.

PREDATORY GIANT OF THE PERMIAN

Above: Reconstruction of a huge semi-aquatic temnospondyl amphibian, *Eryops*, in its Permian environment, based on fossil material from the Red Beds of Texas. It was a common top carnivore in its day and most likely fed on fish and smaller tetrapods. On display at Museum of Ancient Life/Thanksgiving Point Institute, Lehi, Utah, USA. Diorama setting by Chase Studio.

the first turtles, and others. With the rise of aquatic reptiles, marine and freshwater amphibian populations experienced another decline, which resulted in the extinction of all large temnospondyls before the end of the Mesozoic Era.

Sometime between the middle Permian and early Triassic periods, at least one amphibian lineage or perhaps several led to modern forms (**lissamphibians**), possibly derived from small temnospondyls. Paleontologists continue to debate the origins of today's frogs, toads, salamanders, and caecelians, but they were definitely moving among the feet of dinosaurs early in the Jurassic, about 180 mya. Presently the largest living amphibians are the Chinese and Japanese giant salamanders, about 6.5 feet (2 m) in length and up to 130 pounds (59 kg) in weight, and a tropical snake-like caecilian that is about 5 feet (1.5 m) long.

Fossils can offer clues about the behavior of early amphibians, such as where they lived, how they walked or swam, and sometimes what they ate, but for now, details of their natural histories remain a mystery. Nevertheless, we can stir our imagination by looking at how modern amphibians live. Most have lungs and can breathe through their skin, but some have lost their lungs. Most have external fertilization, but some fertilize their eggs internally, like reptiles. Most amphibian females choose a mate and pair up for breeding, but aggressive unpaired males may compete to fertilize a share of the freshly laid eggs. Some

frogs whip up foam nests to protect their developing eggs, while others stick their eggs to leaves and guard them against insect predators. Some have elaborate mating rituals and brood their eggs in strange places—inside dad's mouth or buried in thick skin on mom's back, for example. Some moms give birth to live young in underground hollows and shed fat-rich skin to feed them for weeks. Some larvae are cannibalistic and eat their slower-growing brothers or sisters. Variations on these themes seem endless in the amazing world of amphibians. They are far from being the "simple" creatures most of us think they are.

Of the 6,600 or so known species of living amphibians, about one third are now in serious trouble. Populations are dropping from habitat destruction, introduced species, infectious diseases, pollution, and changes in our planet's climate and atmosphere. Since 1980 biologists have noted that populations of 43% of amphibian species are declining, while only 1% have shown an increase in numbers. Because of their thin, permeable skin and sensitivity to environmental change, amphibians are our planet's "canaries in the coalmine." What is happening to them is a red flag, a sure sign for us to sit up and take notice.

Complex Life Cycles

The origin of modern amphibians and their metamorphosis from free-swimming larvae to adulthood appears to be rooted somewhere in the Paleozoic Era—a picture that paleontologists are still trying to resolve. Fresh insights are coming from rich fossil beds of small, well preserved, salamander-like temnospondyls discovered in Germany—*Apateon* in all stages of development. Some had matured in a larval state—**neoteny**—keeping their aquatic lifestyle, with gills and tail fins. Neoteny is also common among living salamanders—under the right conditions, metamorphosis into terrestrial adults can be delayed or completely eliminated. The larva of one *Apateon* was even found in the digestive track of another of the same species—the first fossil evidence of cannibalism by a larval amphibian.

FISH OUT OF WATER

Bottom: Assemblage of fossil amphibians and fish in mudstone of Permian age, 275 million years old. The large amphibian in the lower left is *Micromelerpeton* and the smaller ones with triangular heads are *Apateon*—mostly 3.5 inches or less in length. Specimen courtesy of Martin Goerlich/Eurofossils.

Cannibalism among living larval salamanders, frogs, and toads has been well documented. Shown here, for example, are two tadpole morphs of the Mexican spadefoot toad, *Spea multiplicata*. Predatory morphs feed on their omnivorous brothers and sisters— with extra protein, the cannibals grow faster and have a better chance of making it to metamorphosis before rain ponds dry up. Captured in the Sonoran Desert, SE Arizona.

Left: Larval tiger salamander from a pond in southern New Mexico will lose its gills and tail fin during metamorphosis. It then relies on lungs and cryptic coloration for life on land and will return to water for breeding.

NOT A ROADKILL

This is an exceptionally well-preserved spadefoot toad, *Eopelobates wagneri*, from Germany's Messel Fossil Site—even its soft parts are visible. In the Messel Pit, 48-million-year-old Eocene oil shale sediments contain a wealth of spectacularly preserved vertebrate animals, insects, and plants. Specimen from a private collection curated by Interprospekt.

PARENTAL CARE

Tropical frogs are full of surprises. Many lay their eggs on leaves or nest on land to avoid aquatic predators. Reticulated glassfrogs (*Hyalinobatrachium valerio*) have two forms of parental care. After mom deposits her sticky egg mass on a leaf over a stream, dad defends them against insect predators, such as ants, grasshoppers, and wasps—day and night. And during the dry season, dad sits on the eggs to protect them from desiccation. Males may continue to call while egg-guarding to entice other females to lay nearby. When the eggs hatch, the tadpoles fall into the water below and stay there through metamorphosis.

The red-eyed treefrog, *Agalychnis callidryas*, is a tropical frog that hangs its eggs on leaves over water. But before laying, mom—with dad riding on her back—makes a trip to the pond to fill her urinary bladder with water. Mom then climbs a nearby plant to release her eggs and water while dad fertilizes them. And sometimes, other males will pile on to add their sperm to the mix. The male shown here is sitting on a torch ginger flower.

Both frogs photographed in Rainmaker Rainforest Reserve, Costa Rica.

The eggs of many animals are soft and decompose quickly, so there is little chance they will fossilize. Nevertheless, some have turned up in the fossil record. Well-preserved eggs of extinct cephalopod mollusks (relatives of squids and the chambered nautilus) have been discovered in 155 million-year-old Jurassic clay in England. Most amphibian eggs look so much like fish eggs as fossils, they are difficult to distinguish from each other, unless of course they are found within the mother's body. Many sharks give birth to live young, but those laying eggs package them in leathery pouches that resist rapid decay. Fossilized egg cases are well known from Mesozoic deposits in the USA, Australia, Germany, and the central Asian republic of Kyrgyzstan. One shark egg case found in England dates back to the Carboniferous, 310 mya.

Most fish, amphibians, and reptiles are **oviparous**—they lay eggs that develop and hatch outside the mother's body. But some species, like great white sharks, guppies, seahorses, fire salamanders, rattlesnakes, and a few chameleons, have an **ovoviparous** mode of reproduction. They carry their developing embryos inside the body, nourish them mainly with egg yolk, and then give birth to live young.

With the advent of the **amniotic egg**, reptiles accelerated the spread of vertebrate life on land. Amniotic eggs are relatively large and well endowed with yolk. Large hatchlings, in turn, are often better competitors and less vulnerable to environmental extremes. To achieve this, the egg must contain not only a rich food supply and a protective outer shell, but also internal chambers, fluids, and a vascular system to aid in the delivery of nutrients, gas exchange, and the elimination of metabolic wastes. Several important membranes make this possible (see diagram).

In amniotic eggs, four membranes provide the embryo's life-support system. The fluid-filled **amnion** cushions the embryo. The **allantois** helps in respiration and stores liquid waste. The yolk sac contains the embryo's primary food supply. And the outermost membrane, the **chorion**, allows the exchange of oxygen and carbon dioxide. The fluid between the chorion and the eggshell is **albumen**—the egg white—which provides the main source of water and some important nutrients for the developing embryo.

Amniote eggshells may be soft or hard. They offer a barrier against water loss and a defense against bacteria, but are porous enough to allow gas exchange. Eggs laid by early reptiles about 340 mya probably had soft shells that preserved poorly, and so far, no Carboniferous-age amniotic eggs have been found. But recently, 280-million-year-old embryos of mesosaurs, which bear a superficial resemblance to crocodiles, were discovered in Uruguay and Brazil. One embryo was curled as if within an egg, and the other was lodged in the remains of a gravid (pregnant) female—their eggs might have been soft-shelled, or perhaps mesosaurs were ovoviparous.

In every geological period since the beginning of the Triassic (252 mya), amniotic eggs with rigid shells have been found worldwide, except in Antarctica. All known crocodiles and birds lay hard-shelled eggs. The first eggshell fragments of puzzlingly big eggs were described by a Catholic Priest in France in 1859. At that time scientists knew little about dinosaurs, so the dinosaurian origin of these eggshells remained a mystery for many years. In 1923, a field team from the American Museum of Natural History (AMNH) found dino eggs alongside dinosaur bones in Mongolia. And since the 1980s, many hard-shelled dinosaur eggs have been surfacing on other continents.

Then in 2020, news of soft-shelled dinosaur eggs hit the headlines. Using LSF imaging (see Chapter 8), AMNH paleontologist Mark Norell and a team of eight colleagues studied two unusual clutches of dino eggs and found clear evidence of soft shells. The *Mussaurus* clutch from Argentina was laid about 215 mya, and the *Protoceratops* eggs from Mongolia about 75 mya. So it's now clear that some dinosaurs were laying soft-shelled eggs throughout the Mesozoic Era.

Most dinosaur eggs are about the size of a grapefruit, but some are as small as goose eggs. They vary in shape from spherical to oblong, the largest known being about 2 feet (60 cm) in length, and about 8 inches (20 cm) wide. Yet by volume, eggs of the recently extinct elephant bird hold the size record—equivalent to 150 chicken eggs or six ostrich eggs! Although you might expect titanosaurs to lay enormous eggs, bigger eggs can be a handicap. Structurally, they require thicker shells for support, and thicker shells impede respiration and hatching. Furthermore, in environments with heavy predation on the young, females that reach sexual maturity quickly and lay smaller eggs but more of them can hold a survival advantage—a trend evident in modern populations of iguanas and birds.

Mammals are nearly all **viviparous**—they give birth to live young that have been nourished not by egg yolk but solely by the mother's tissues. Mammals are considered amniotes because their embryos develop within an amniotic sac. The other three membranes are present too, but in modified form. The yolk sac and allantois have become the fetal **umbilical cord**, which delivers food to the embryo and removes wastes. These membranes, plus part of the chorion, form the **placenta**, a vascular organ that unites the embryo with the mother's uterus. Food, air, and wastes are transferred across the placenta. As in amniotic eggs, the chorion surrounds the embryo and its membranes, and when the fluid-filled chorion "breaks," birthing begins. The duck-billed platypus and echidnas—**monotreme mammals**—are the only living exceptions. They lay eggs but share other key anatomical features possessed by placental mammals: for example, they are hairy and have mammary glands that secrete milk for their babies.

Chorion membrane

Big & Small, Old & New

Left: A nest of dinosaur eggs—possibly from a therizinosaur (= segnosaur)—of Cretaceous age (146–65 mya), collected in Nanyang County, Henan Province, China. These eggs are the size of a navel orange, smaller than an ostrich egg. Specimen courtesy of geologist Zhouping Guo.

Above: Model of an oviraptor dinosaur egg with embryo about to hatch. Based on a specimen of Late Cretaceous age from Mongolia. Artwork by Dennis Wilson/Pangea Designs.

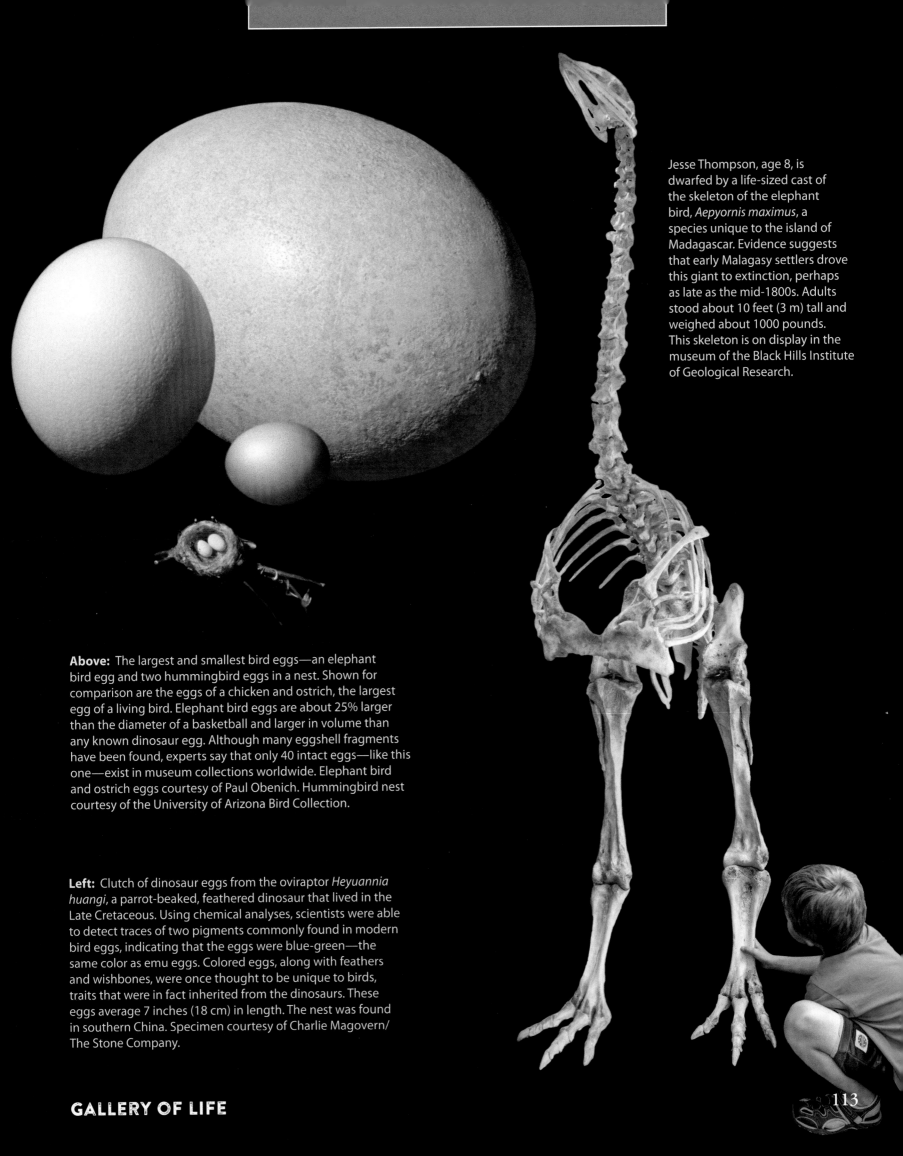

Jesse Thompson, age 8, is dwarfed by a life-sized cast of the skeleton of the elephant bird, *Aepyornis maximus*, a species unique to the island of Madagascar. Evidence suggests that early Malagasy settlers drove this giant to extinction, perhaps as late as the mid-1800s. Adults stood about 10 feet (3 m) tall and weighed about 1000 pounds. This skeleton is on display in the museum of the Black Hills Institute of Geological Research.

Above: The largest and smallest bird eggs—an elephant bird egg and two hummingbird eggs in a nest. Shown for comparison are the eggs of a chicken and ostrich, the largest egg of a living bird. Elephant bird eggs are about 25% larger than the diameter of a basketball and larger in volume than any known dinosaur egg. Although many eggshell fragments have been found, experts say that only 40 intact eggs—like this one—exist in museum collections worldwide. Elephant bird and ostrich eggs courtesy of Paul Obenich. Hummingbird nest courtesy of the University of Arizona Bird Collection.

Left: Clutch of dinosaur eggs from the oviraptor *Heyuannia huangi*, a parrot-beaked, feathered dinosaur that lived in the Late Cretaceous. Using chemical analyses, scientists were able to detect traces of two pigments commonly found in modern bird eggs, indicating that the eggs were blue-green—the same color as emu eggs. Colored eggs, along with feathers and wishbones, were once thought to be unique to birds, traits that were in fact inherited from the dinosaurs. These eggs average 7 inches (18 cm) in length. The nest was found in southern China. Specimen courtesy of Charlie Magovern/ The Stone Company.

Late in the Paleozoic Era, plant life expanded onto the super-continent Pangea, which was divided by huge slow-moving rivers. Land animals followed plant life and flourished. In marshes of ferns and club mosses lived giant amphibians—six feet of sluggishness with huge, toothy, spade-shaped heads and long tails. From amphibian ancestors came the amniotes, bringing a diversity of terrestrial life-styles. These newer lineages include reptilian predecessors of the archosaurs (dinosaurs, crocodiles, and birds); primitive turtles; "stem reptiles" from which most other reptiles are thought to have evolved; and proto-mammals, known as therapsids.

Many of these creatures disappeared during Earth's third mass extinction event—the **Permian extinction**—the biggest ever, which ended the Paleozoic Era 252 mya. The largest land-lubbers vanished. Fortunately, a few reptilian and proto-mammalian lineages survived into the Mesozoic Era—the Age of Reptiles. The **archosaurs** achieved dominance early in the Mesozoic, developing into a lineage of croc-like reptiles from which dinosaurs, pterosaurs, birds, and crocodiles evolved. As Pangea began to break up at this time, populations of reptilian and mammalian ancestors became fragmented and isolated, favoring the evolution of new species.

During the Mesozoic Era—the Age of Reptiles—dinosaurs flourished on land while other reptiles swam into fame as top predators of the sea. All had lungs, so to breathe they had to surface, like today's turtles and marine mammals. Dolphin-like *ichthyosaurs* (Greek for "fish-lizard")—shown above—shared the ocean with giant snake-necked *plesiosaurs*—below. This 180 million-year-old ichthyosaur, *Stenopterygius*—courtesy of Martin Goerlich—was found in Holzmaden shale in southern Germany. Dark remains of its stomach contents can be seen within its rib cage, and rare specimens show ichthyosaurs giving birth.

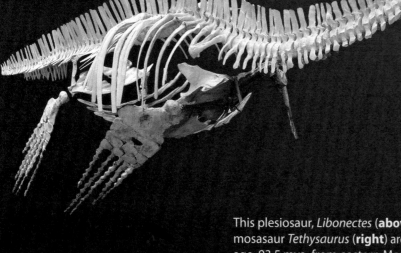

This plesiosaur, *Libonectes* (**above**), and the primitive mosasaur *Tethysaurus* (**right**) are fossils of the same age, 93.5 mya, from eastern Morocco. Plesiosaurs were well established as marine predators at this time, but mosasaurs were still transitioning from a four-legged anatomy to a sea-worthy design with paddle-like flippers and a long, flattened tail. Photographs and specimens—both from eastern Morocco—courtesy of Serge Xerri.

Pterygoid teeth
on roof of mouth
(from a different
mosasaur skull)

While dinosaurs walked their way into the Mesozoic, other reptiles rose to power in the oceans. In spite of their need to breathe air, more than a dozen groups of sea-faring reptiles appeared in the fossil record at this time. Most of the smaller groups disappeared early in the Mesozoic at the end of the Triassic Period (252–201 mya)—among them were pachypleurosaurs, nothosaurs, placodonts, and thalattosaurs—a bizarre array of mostly predators. Two other groups that got their start in the Triassic continued to thrive through the Mesozoic: the dolphin-like **ichthyosaurs** and paddling **plesiosaurs**—those famous barrel-bodied fish predators with small heads and long, snake-like necks. Massive-headed, short-necked cousins of plesiosaurs, the **pliosaurs**, arrived in the middle of the Mesozoic (during the Jurassic).

A ferocious group of late-comers, the **mosasaurs**, held command of the sea for at least 24 million years late in the Mesozoic, about 90–66 mya. Fossils of these sleek, fast-swimming, toothy predators have revealed that they gave birth to live young (as ichthyosaurs did) and ate everything from ammonites and fish to plesiosaurs and diving seabirds. In oceans worldwide, sea turtles and some crocs were the only marine reptiles that survived the cataclysm that brought the glory days of non-avian dinosaurs to an abrupt end.

MOSASAURS

Mosasaur fossils are spectacular, abundant, diverse, and well-studied. Some species were longer than a city bus, a few were smaller than an alligator, but most were somewhere in between. This 70 million-year-old skeleton of *Mosasaurus beaugei* (**above**) was collected, prepared, and photographed in Morocco by Serge Xerri.

Like modern lizards and snakes, these extinct reptiles had scaly skin. And their skulls share ancestral features with today's snakes and some lizards. Mosasaurs have a rigid upper jaw like that of lizards, but their double-hinged lower jaw is more snake-like. A joint in the lower jaw gives it flexibility for engulfing large prey. Snakes have loosely hinged bones in both upper and lower jaws, allowing them to stretch their mouths to enormous proportions. Most snakes and all mosasaurs also have teeth in the roof of the mouth (*pterygoid teeth*) that help to grip their victim while being swallowed whole.

Ammonites with puncture holes are sometimes found in fossil beds with mosasaur bones. The size, shape, and bite pattern of these holes strongly suggest that ammonites were on the mosasaurian menu. In John Fischner's fanciful sculpture "Revenge of the Ammonites" (**right**), he has turned prey into predator for some comic relief

Background photo by Barry B. Brown

The prehistory of modern lizards and snakes, the **squamates**, and their sister group, the **sphenodontids**, has been challenging to piece together. Squamate fossils have been small and scarce, giving paleontologists little to work with. One mystery fossil from the Italian Alps, which sat in a museum collection for 20 years, was recently re-examined using powerful non-invasive CT X-ray scanning technology (see Chapter 8, *Seeing the Unseen*). This specimen turned out to be a 240-million-year-old lizard, the oldest known. There is only one living sphenodontid, the **tuatara**, and it is unique to New Zealand. The tuatara's family line dates back to the Triassic, also about 240 mya, and no less than 16 extinct species in this family have been described. So we now know that both groups originated early in the Mesozoic Era, in the shadow of dinosaurs.

Throughout their history, **turtles** were on a separate evolutionary path, which is still being reconstructed as new fossils are discovered. The development of the turtle's shell has puzzled paleontologists for centuries. The shell's anatomy suggests that it originated from a wide backbone with flattened, fused ribs. One promising ancestor, *Eunotosaurus*, had flat elongated ribs curved over its back; it appeared in the Permian (299–252 mya) of the Paleozoic. And recent fossil finds of intermediate forms in China and Germany have placed the evolution of the shell firmly in the Triassic Period. Turtles known from Jurassic times, about 200 mya, share the same unmistakable body plan seen in today's turtles. Some of these extinct marine species, *Archelon* for example, were twice the size of the leatherback sea turtle, our largest living species.

TWO-LEGGED WORM LIZARD

Based on their many shared anatomical features, we know that snakes evolved from lizards, but with so few fossils to work with, reconstructing their past has been slow. Different lineages of modern lizards—especially those living underground—have evolved snake-like body forms with or without legs. But unlike snakes, burrowing lizards have kept relatively inflexible skulls. Consider Baja California's two-legged worm lizard, *Bipes*, for example. It is unusual in having mole-like front legs for digging and no hindlimbs (**above**).

Above: Snakes found in Early Eocene (56–48 mya) deposits all have a modern body design. This cast of an extremely rare, well-preserved boa, *Boavus idelmani*, from the Green River Formation in Wyoming, is on exhibit at Fossil Butte National Monument. The original is housed at the Houston Museum of Natural Science.

Mounting evidence suggests at least some snakes evolved from small, elongated, sea-going lizards with reduced limbs no less than 120 million years ago. Other fossil evidence points to a long early history of snakes with hindlimbs and no front legs. And genetic studies of modern snakes show kinship with monitor lizards—the Komodo dragon, for example, the world's largest living lizard (**below**). A monitor's forked, snake-like tongue provides a sense of smell in stereo. Each tip "tastes" the air for chemical cues in its environment. Photographed at St. Augustine Alligator Farm, Florida.

Above: Endowed with amniotic eggs, early reptilian life forms began to diversify late in the Paleozoic Era. Among the pioneers was *Claudiosaurus*, a lizard-like creature that lived along the shores of lakes and sheltered seas in Permian times (299–252 mya). Fossils suggest that it was a good swimmer, much like today's marine iguana, and it vanished 252 mya during the great Permian extinction. This 17-inch (43-cm) specimen—courtesy of Stefan & Christine Perner/Earth Art Gallery—is from Madagascar.

"ALIEN" SEA TURTLE

Four-year-old Ryan Christiansen, a visitor at Tucson's Gem, Mineral, and Fossil Showcase, is awe-struck by this new species of giant sea turtle, *Alienochelys selloumi*, of Late Cretaceous age (70–66 mya), from Khouribga Province, Morocco. This unusual turtle has a wide, flaring beak and a sturdy skull with shell-crushing plates. This specimen measures 9 feet (2.7 m) in length—courtesy of Serge Xerri from Rabat, Morocco.

Tortoises (family Testudinidae) are turtles that are well adapted for life on land, so all tortoises are turtles, but not all turtles are tortoises. Starting early in the Cenozoic Era (66 mya to the present), tortoises crawled into the fossil record. Extinct and living species are known from North and South America, Europe, Africa, Asia, and many islands in warm parts of the world. Most have a domed shell and some attained gigantic body size—the largest known was *Megalochelys (=Colossochelys)*, a one-ton tank from southern Asia. Esmeralda, a male Aldabra tortoise that's roaming freely on an island in the Seychelles, holds a place in the *Guinness Book of World Records*—he is the heaviest living tortoise in the wild. In 2002 he weighed 800 pounds (363 kg), and he's now about 175 years old. A captive Aldabra tortoise named Adwaita that was gifted to a British general in the 1700s and later transferred to India's Alipore Zoo in 1875 died there in 2006. Carbon-dating its shell confirmed that this tortoise was born around 1750 and had lived for about 255 years—the longest documented life span for any animal on Earth!

What are the advantages of gigantic body size in terrestrial creatures like giant tortoises and the biggest dinosaurs, the titanosaurs? Most obvious is protection against predators. Larger animals also have slower metabolism. And in non-avian reptiles, reduced energy consumption correlates with a longer life span. Perhaps most important, all of these giants were and still are herbivorous. Food spends more time in their specialized gut, which permits a more efficient use of plant foods of lower nutritional value.

FRESHWATER TURTLE

This beautifully preserved soft-shelled turtle, *Trionyx sp.*, of Eocene age (about 50 mya) is one of the largest ever found in Wyoming's Green River fossil beds—5 feet 10 inches (1.8 m) from snout to tip of its tail. Specimen courtesy of Jim Tynsky; prepared by Dan Ulmer.

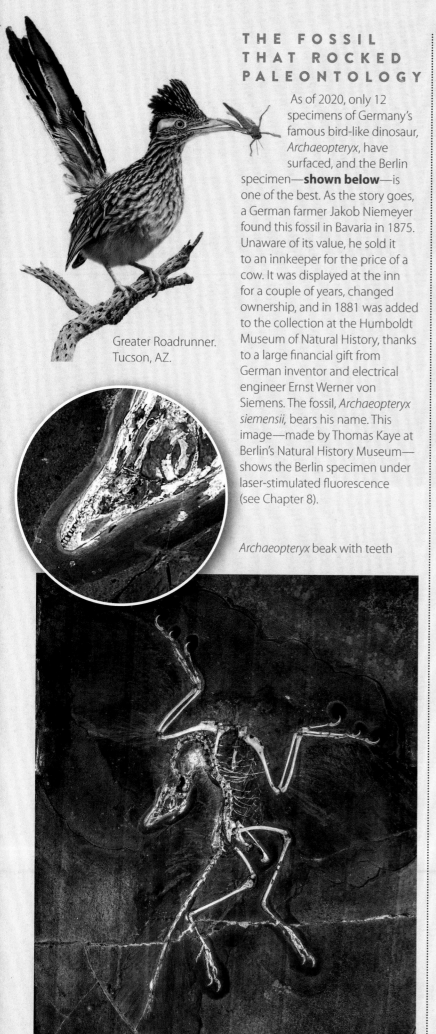

THE FOSSIL THAT ROCKED PALEONTOLOGY

As of 2020, only 12 specimens of Germany's famous bird-like dinosaur, *Archaeopteryx*, have surfaced, and the Berlin specimen—**shown below**—is one of the best. As the story goes, a German farmer Jakob Niemeyer found this fossil in Bavaria in 1875. Unaware of its value, he sold it to an innkeeper for the price of a cow. It was displayed at the inn for a couple of years, changed ownership, and in 1881 was added to the collection at the Humboldt Museum of Natural History, thanks to a large financial gift from German inventor and electrical engineer Ernst Werner von Siemens. The fossil, *Archaeopteryx siemensii*, bears his name. This image—made by Thomas Kaye at Berlin's Natural History Museum—shows the Berlin specimen under laser-stimulated fluorescence (see Chapter 8).

Greater Roadrunner. Tucson, AZ.

Archaeopteryx beak with teeth

Among the archosaurs, the biggest and fiercest dinosaurs always get the most attention. In fact, more books have been written about them than any other group of prehistoric animals. But in the bigger picture, dinosaurs that never attained a great size were more significant in the evolution of life on Earth. The innovation of flight came not from the giants but from smaller forms that were easy to overlook.

In 1859, with the publication of *On the Origin of Species*, Charles Darwin predicted that paleontologists would someday discover a fossil that provided evidence of a transition between reptiles and modern birds. And only two years later, the first complete skeleton of *Archaeopteryx* was found in Germany—a 150-million-year-old, roadrunner-sized dinosaur with hollow, thin-walled bones, feathers, a wishbone, and wings. This fossil also shares unmistakable features with theropod dinosaurs—a "beak" with sharp teeth, a long bony tail, abdominal ribs, and three digits on each "hand." But could *Archaeopteryx* fly? Recent CT X-ray scans of its skeletal architecture have revealed that this dino-bird was indeed capable of short bursts of active flight, much like modern roadrunners.

A more complete picture of the **dinosaur-bird transition** surfaced in China's Liaoning Province during the 1990s. Miraculous treasure troves of Jurassic and Cretaceous fossils were discovered in fine-grained lake deposits of volcanic ash. The victims had been quickly entombed under oxygen-free conditions, which prevented decay and preserved their bones, feathers, skin, internal organs, and sometimes stomach contents in exquisite detail. A rich abundance of life forms—from plants and snails to no less than 53 species of prehistoric birds—have been uncovered in Liaoning, with more being unearthed every year.

With so many new fossils to study, paleontologists now have a clearer picture of how and when some dinosaurs became birds. Feathers and flight evolved independently multiple times in the Dinosauria. *Yutyrannus*, for example, a 3,100-pound (1,400-kg) feathered dinosaur related to the infamous *T. rex*, probably used its simple feathers not for flight but for display and to corral and capture prey. Only some of the smaller species acquired more specialized anatomical features that made the evolutionary transition to free-flight possible. Armed with a raft of new discoveries, mostly coming from China, the scientific community now agrees that modern birds are the direct descendants of dinosaurs. Remarkable finds of dinosaur nests, embryos, and hatchlings are also offering a patchy picture of dinosaurs as creatures with parental care and colonial nesting.

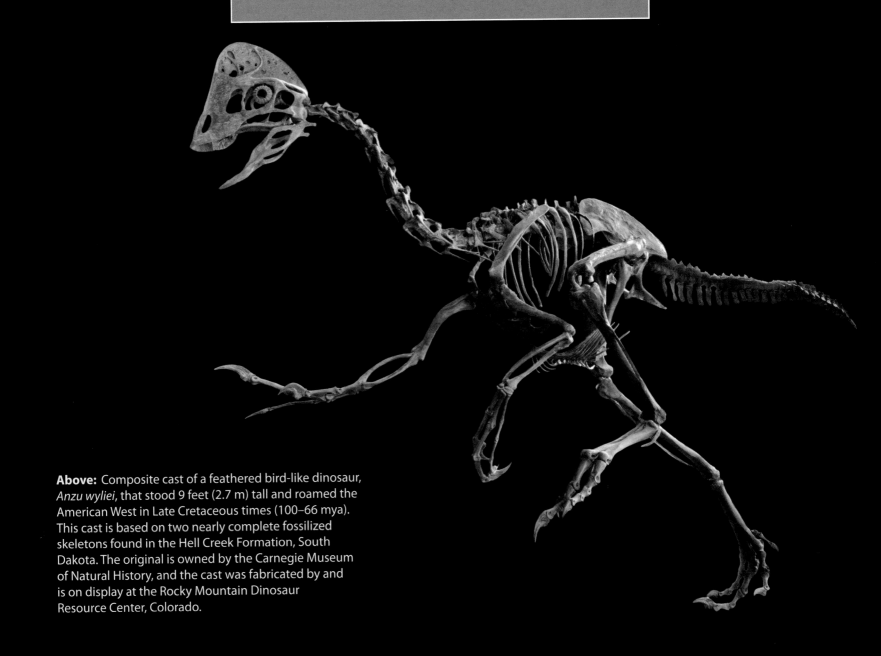

Above: Composite cast of a feathered bird-like dinosaur, *Anzu wyliei*, that stood 9 feet (2.7 m) tall and roamed the American West in Late Cretaceous times (100–66 mya). This cast is based on two nearly complete fossilized skeletons found in the Hell Creek Formation, South Dakota. The original is owned by the Carnegie Museum of Natural History, and the cast was fabricated by and is on display at the Rocky Mountain Dinosaur Resource Center, Colorado.

Left: *Confuciusornis sanctus,* a primitive bird with a toothless beak. It had claws on its wings, a character derived from its therapod dinosaur ancestors that is missing in most modern birds—the hoatzin is an exception. Hundreds of specimens of this bird have been recovered from Early Cretaceous (145–100 mya) lake deposits in Liaoning Province, northeastern China. This specimen is 27 inches (69 cm) from the tip of its beak to the tip of the longest tail feather; it's housed at the Black Hills Institute of Geological Research.

While dino-birds were attempting to perfect flight, other reptiles soared overhead, the **pterosaurs** (commonly, but not accurately, called "pterodactyls"—pterodactyls represent only one species group of pterosaurs). These masters of Mesozoic skies were already flourishing when dinosaurs and sea-faring reptiles were just coming into power about 230–200 mya. They are not closely related to birds, or bats, and contrary to popular belief, they are not dinosaurs. Pterosaurs blazed an independent path to success without flight feathers. To lighten the load for flight, pterosaur bones were hollow and not easily preserved as fossils. Consequently, early pterosaur fossils are scarce, and the origin of this amazing lineage remains unclear. All of them died out toward the end of the Cretaceous, along with their earth-bound dinosaur cousins, possibly from competition with birds.

Worldwide, new pterosaur fossils are being found every year, allowing paleontologists to fill in some of the blanks. More than 110 species have now been described. A cache of 215 two-inch (5-cm) eggs found in China suggests colonial nesting. Some species were covered in hair-like feathers, so they might have been warm-blooded. Many evolved bizarre and specialized head shapes, with or without teeth. As in dinosaurs, some pterosaurs

were small and their lineage moved toward gigantism in the Cretaceous. The largest known flying animal, *Quetzalcoatlus northropi*, found in Texas, was as tall as a giraffe and had a wingspan of about 33 feet (12 m), matching that of an F-16 fighter jet!

As a final note, the classification of reptiles is in a constant state of change. With the discovery of new fossils and the application of more modern analytical techniques, paleontologists often rearrange species groups and change their names. Paleontologists also have differing philosophies about how these groups should be organized to reflect our best understanding of reptilian evolution.

PTEROSAUR

Prehistoric flying reptile, a short-tailed pterosaur, *Rhamphorhynchus kochi*—surrounded by colorful mineral dendrites—in Solnhofen limestone of Late Jurassic age (164–145 mya), from Bavaria, Germany. Its skull is 3.5 inches (8.8 cm) in length. Specimen courtesy of Raimund Albersdoerfer, Germany.

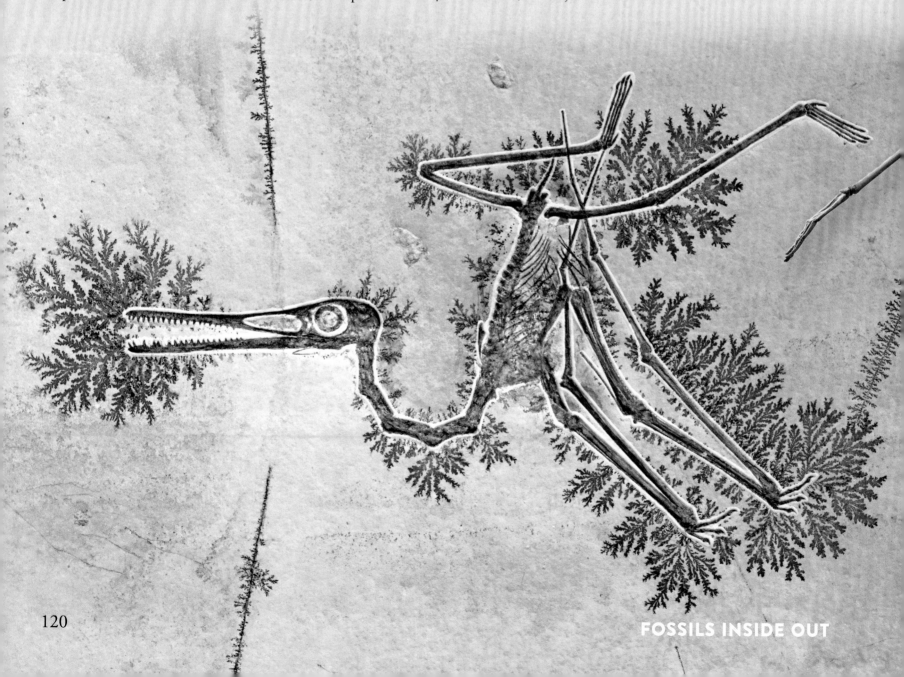

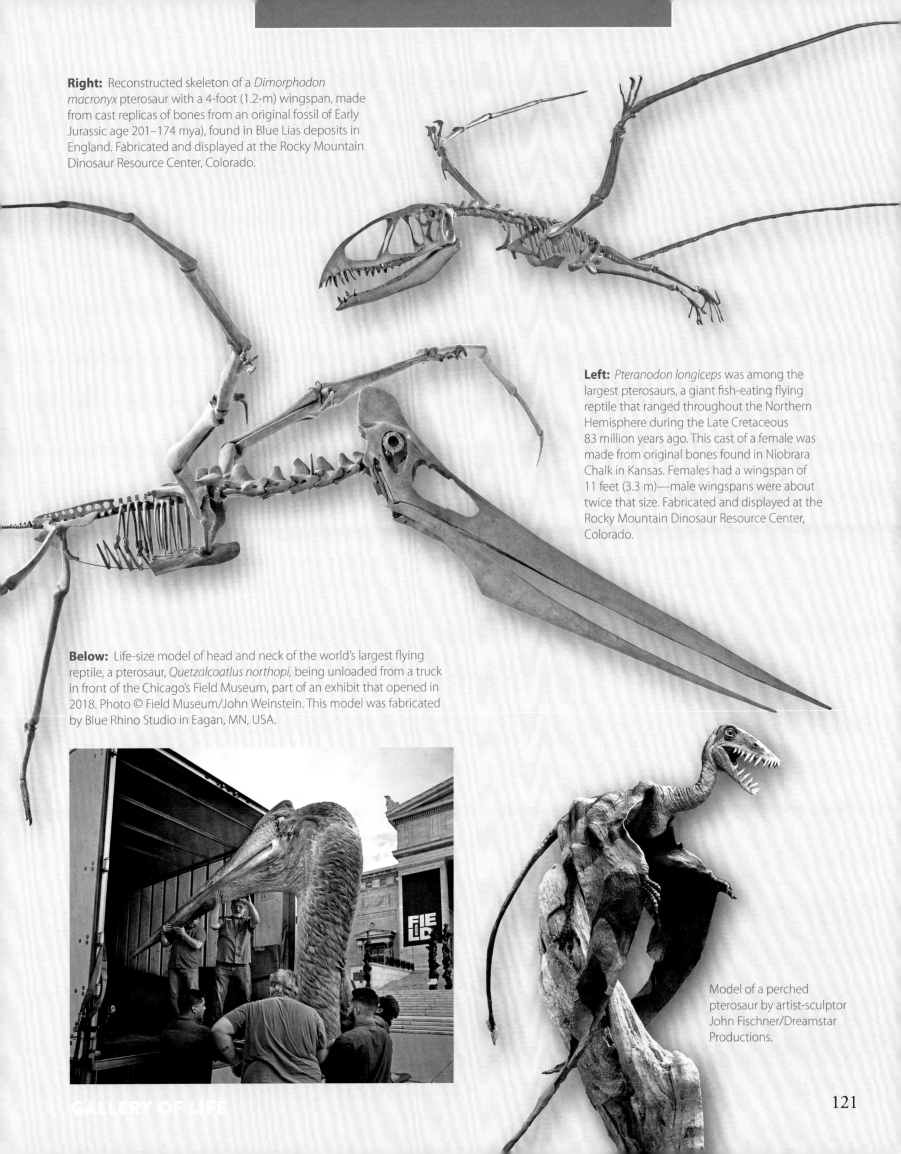

Right: Reconstructed skeleton of a *Dimorphodon macronyx* pterosaur with a 4-foot (1.2-m) wingspan, made from cast replicas of bones from an original fossil of Early Jurassic age 201–174 mya), found in Blue Lias deposits in England. Fabricated and displayed at the Rocky Mountain Dinosaur Resource Center, Colorado.

Left: *Pteranodon longiceps* was among the largest pterosaurs, a giant fish-eating flying reptile that ranged throughout the Northern Hemisphere during the Late Cretaceous 83 million years ago. This cast of a female was made from original bones found in Niobrara Chalk in Kansas. Females had a wingspan of 11 feet (3.3 m)—male wingspans were about twice that size. Fabricated and displayed at the Rocky Mountain Dinosaur Resource Center, Colorado.

Below: Life-size model of head and neck of the world's largest flying reptile, a pterosaur, *Quetzalcoatlus northopi,* being unloaded from a truck in front of the Chicago's Field Museum, part of an exhibit that opened in 2018. Photo © Field Museum/John Weinstein. This model was fabricated by Blue Rhino Studio in Eagan, MN, USA.

Model of a perched pterosaur by artist-sculptor John Fischner/Dreamstar Productions.

RISE OF MAMMALS

The demise of dinosaurian giants ushered in the Age of Mammals, the **Cenozoic** (or *recent life*) **Era**, which covers the last 66 million years of Earth's history. Although mammals made a relatively inconspicuous debut at the feet of dinosaurs during the Mesozoic Era, their triumphant entry into the fossil record began 66–56mya (see *Calendar of Life* in Chapter 4). Ecological niches left empty by massive dinosaur die-offs were soon filled with an astonishing diversity of new mammals, along with birds that descended from a lineage of dinosaurs. In just 10 million years— the blink of an eye in geological time—thousands of mammal species evolved, including whales, bats, and primates.

All mammals have mammary glands that make milk to nourish their young—the origin of the name *mammal*. But mammary glands and other distinguishing soft parts rarely preserve well. Mammalian features easier to identify in the fossil record are hair, two characteristic bones in the middle ear (*malleus* and *incus*), the unique hinging mechanism of the lower jaw, and their revolutionary dentition. In fact, the success of mammals has much to do with their **teeth**. In different groups of mammals, teeth became highly specialized in form and function—for piercing, cutting, tearing, crushing, grinding, digging, and gripping.

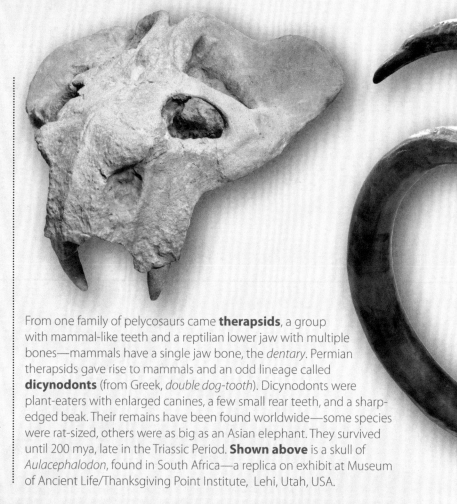

From one family of pelycosaurs came **therapsids**, a group with mammal-like teeth and a reptilian lower jaw with multiple bones—mammals have a single jaw bone, the *dentary*. Permian therapsids gave rise to mammals and an odd lineage called **dicynodonts** (from Greek, *double dog-tooth*). Dicynodonts were plant-eaters with enlarged canines, a few small rear teeth, and a sharp-edged beak. Their remains have been found worldwide—some species were rat-sized, others were as big as an Asian elephant. They survived until 200 mya, late in the Triassic Period. **Shown above** is a skull of *Aulacephalodon*, found in South Africa—a replica on exhibit at Museum of Ancient Life/Thanksgiving Point Institute, Lehi, Utah, USA.

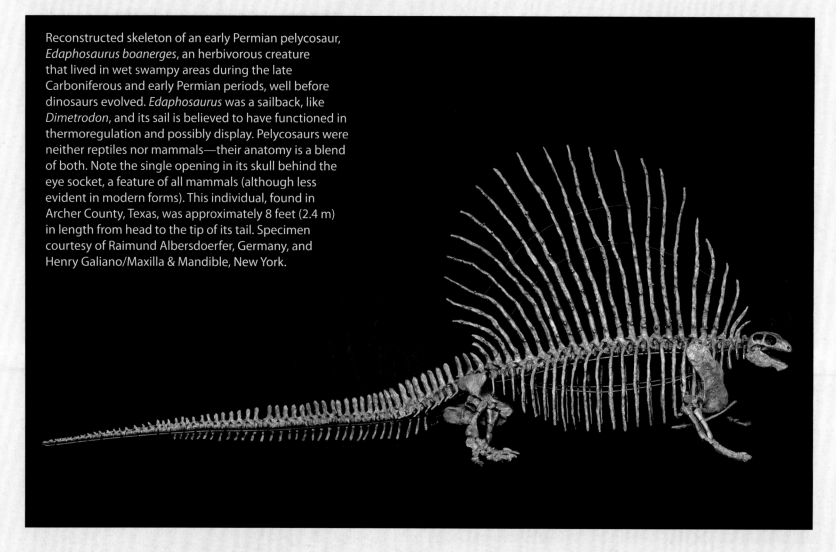

Reconstructed skeleton of an early Permian pelycosaur, *Edaphosaurus boanerges*, an herbivorous creature that lived in wet swampy areas during the late Carboniferous and early Permian periods, well before dinosaurs evolved. *Edaphosaurus* was a sailback, like *Dimetrodon*, and its sail is believed to have functioned in thermoregulation and possibly display. Pelycosaurs were neither reptiles nor mammals—their anatomy is a blend of both. Note the single opening in its skull behind the eye socket, a feature of all mammals (although less evident in modern forms). This individual, found in Archer County, Texas, was approximately 8 feet (2.4 m) in length from head to the tip of its tail. Specimen courtesy of Raimund Albersdoerfer, Germany, and Henry Galiano/Maxilla & Mandible, New York.

SPECIALIZED TEETH

An exact replica of a fossilized woolly mammoth skull from Alaska, USA. Its tusks—enlarged incisor teeth—are 7.5 feet (2.3 m) long; and adults were about the size of today's Indian elephant. Replica courtesy of David Kronen/Bone Clones.

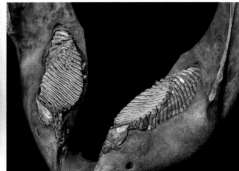

Above: Beautifully preserved lower jaw, the mandible, of a young woolly mammoth. As in modern elephants, mammoths had only four, large, ridged molars in their mouth, well suited to grinding tough plant foods. As these teeth wear down, they are replaced with a new set—up to six times in an elephant's life. This 30,000-year-old Pleistocene specimen is from Siberia, Russia, courtesy of Stefano Piccini/Geoworld Group.

Teeth speak volumes about the lifestyle of their owner. Grassland **herbivores**, such as horses, camels, and deer, are endowed with front teeth for nipping and broad, flat molars for crushing and grinding tough plant foods. Most **carnivores** are equipped with enlarged and pointed canine teeth for gripping prey, along with shearing teeth for cutting and tearing flesh. Their molars are reduced or missing. And **omnivores**—pigs, bears, and humans, for example—eat both plants and animals. Their "hybrid dentition" allows them to cut, tear, and crush different types of food.

Tusks—as seen in elephants and their prehistoric relatives—are enlarged front teeth specialized for digging, defense, and displays of dominance among males. Strangest of all, the solitary unicorn-like tusk of the narwhal grows inside out—its hard tissue is on the inside, with nerves on the outside. Recent drone footage revealed that these whales use their unique, spirally twisted tooth to smack and stun fish. Yet most females have no tusk. This specialized tooth must have a sensory function, among other possibilities, but scientists remain puzzled by it.

Above: Skull of a coyote-sized carnivore, *Hyaenodon crucians*, with exaggerated teeth and jaws. This member of the extinct group of prehistoric mammals Creodonta (meaning *flesh-tooth* in Greek) roamed plains of North America, Eurasia, and Africa in Early to Middle Oligocene times (33.9–28.4 mya). Specimen from the Brule Formation, South Dakota, courtesy of Japheth Boyce Fossils/Rapid City, SD.

Right: Skull showing the teeth of a typical omnivore—a primitive pig-like mammal, *Archeotherium*, a group that lived only in North America and Eurasia during the Eocene, Oligocene, and early Miocene epochs. This specimen was found in the White River Badlands, South Dakota, USA. The skull of this fossil is 19 inches (48 cm) long; courtesy of Japheth Boyce Fossils.

MINIATURE HORSES

About 52–48 mya, early in the Eocene, miniature horses browsed in subtropical woodlands of North America and Eurasia. Superb fossils of these early horses have been found in Wyoming and in Germany's Messel Pit. In life, they were about the size of an English setter and had no hooves—a hoof is a single weight-bearing middle digit with an enlarged toenail. Instead, they had 3-4 "hooflets" on each foot.

Eohippus—meaning "dawn horse," and formerly named *Hyracotherium*—is the earliest known ancestor of the modern horse. It's a primitive "odd-toed ungulate," a group that includes today's horses, zebras, rhinos, and tapirs. The most complete articulated skeleton of this prehistoric horse ever discovered (shown here) was found in 2003 in a private quarry near Kemmerer, Wyoming. This magnificent 50-million-year-old specimen, courtesy of Jim Tynsky, was prepared by Dan Ulmer. Experts at the Chicago Field Museum determined that it was a mature adult.

Early in the Cenozoic Era, during the **Paleogene Period**, 66–23 mya, flora and fauna flourished in and around subtropical lakes, streams, and lagoons, leaving rich deposits of fossils in freshwater and marine sediments. Mammals became firmly established in all major land ecosystems, while some took to the air (bats) and others populated the oceans (whales). Two of the world's most diverse and abundant fossil deposits of this age reside in Germany's **Messel Pit** and North America's **Green River Formation**. Sediments from these ancient freshwater lake basins are time capsules that have given us a remarkable view of life during the **Eocene** (the *dawn of new times* epoch), 56–34 mya. North America, Europe, and Asia were loosely connected land masses back then, so it's not surprising that Eocene fossils from these three continents share many similarities. Along with bats, dog-sized "dawn horses," rat-sized primates, and strange tree-climbing carnivores, these deposits were full of fish, turtles, palm fronds, shorebirds, crocodiles, lizards, snakes, insects—the list goes on and on—enough fossils to allow the reconstruction of complete Eocene ecosystems. Many of these species closely resemble plants and animals familiar to us today.

About 34 mya, at the onset of the **Oligocene Epoch**, global climates became cooler and drier. Tropical and subtropical forests gradually gave way to cold-hardy open woodlands, grasslands, and deserts. The fossil record shows a shuffling of species as the environment changed, and many mammals went extinct, especially in Europe. As savanna-like grasslands expanded, so did the numbers and types of grazers and browsers, including camels, rhinos, tapirs, horses, oreodonts (a group of extinct pig/camel/sheep-like hoofed mammals), and early ancestors of deer and cattle. Among the predators, "wolf-like" *Hyaenodon*, primitive saber-tooth "cats," and "bear-dogs" stalked the herbivores. Few of these early mammals attained the size of a modern horse; but one, the Asian hornless rhino, *Indricotherium*, holds the record for being the largest known land mammal. It was three to four times the weight of an African elephant. In North America, rich fossil deposits of Oligocene mammals have been found in the White River Badlands of Wyoming, South Dakota, Colorado, Nebraska, and Canada.

While there was much migration of mammals between North America, Europe, and Asia at this time, these continents were largely isolated from Africa, Australia, South America, and Antarctica. Only a thin chain of islands "bridged" North and South America, and connections between Eurasia and Africa were equally sparse and sporadic. With episodes of volcanic activity, mountain-building, climate change, and fluctuating sea level 66–23 mya, some groups spread and others vanished. Certain mammals appear to have evolved on continents where they now reside; others appear to be immigrants. Some bizarre giants populated South America, including elephant-like *Pyrotherium* and the meat-eating marsupial *Borhyaena*, which lived alongside huge penguins and flightless, predatory "terror birds."

TERMINATOR PIG

Life-like reconstruction of the biggest of the "terminator pigs," *Daeodon shoshonensis (= Dinohyus hollandi)*. A standing man could stare this beast straight in the eye. This extinct group of pig-like mammals, the entelodonts, inhabited North America 29–19 mya during the late Oligocene and early Miocene epochs. In many respects, these even-toed ungulates are anatomically more similar to hippos and whales than to pigs. This diorama is based on fossilized remains found in Agate Springs, Nebraska. On display in the Prehistoric Journey exhibit at the Denver Museum of Nature and Science in Colorado. Sculpture of the mammal by Tom Shankster; background art by Jeff Wrona.

In situ preparation of an **oreodont** skeleton, *Minochoerus*, of Late Oligocene age, from the Brule Formation in Wyoming, USA. Specimen courtesy of Japh Boyce Fossils.

During the Eocene, **brontotheres** were the largest land mammals to browse in North American and Eurasian forests, some reaching 14 feet (4.3 m) in length. The head was rhinoceros-like, but its bony horns differ from those of rhinos—rhino horns are matted hair without a bony core. The earliest forms were small and hornless. Brontotheres vanished from the fossil record at the close of the Eocene, and although many fossils of this diverse family have surfaced, their place among odd-toed ungulates has remained a puzzle. This nearly complete skull of *Brontops robustus* from the White River Badlands in South Dakota was about 35 million years old, a specimen courtesy of Henry Galiano/Maxilla & Mandible, Ltd.

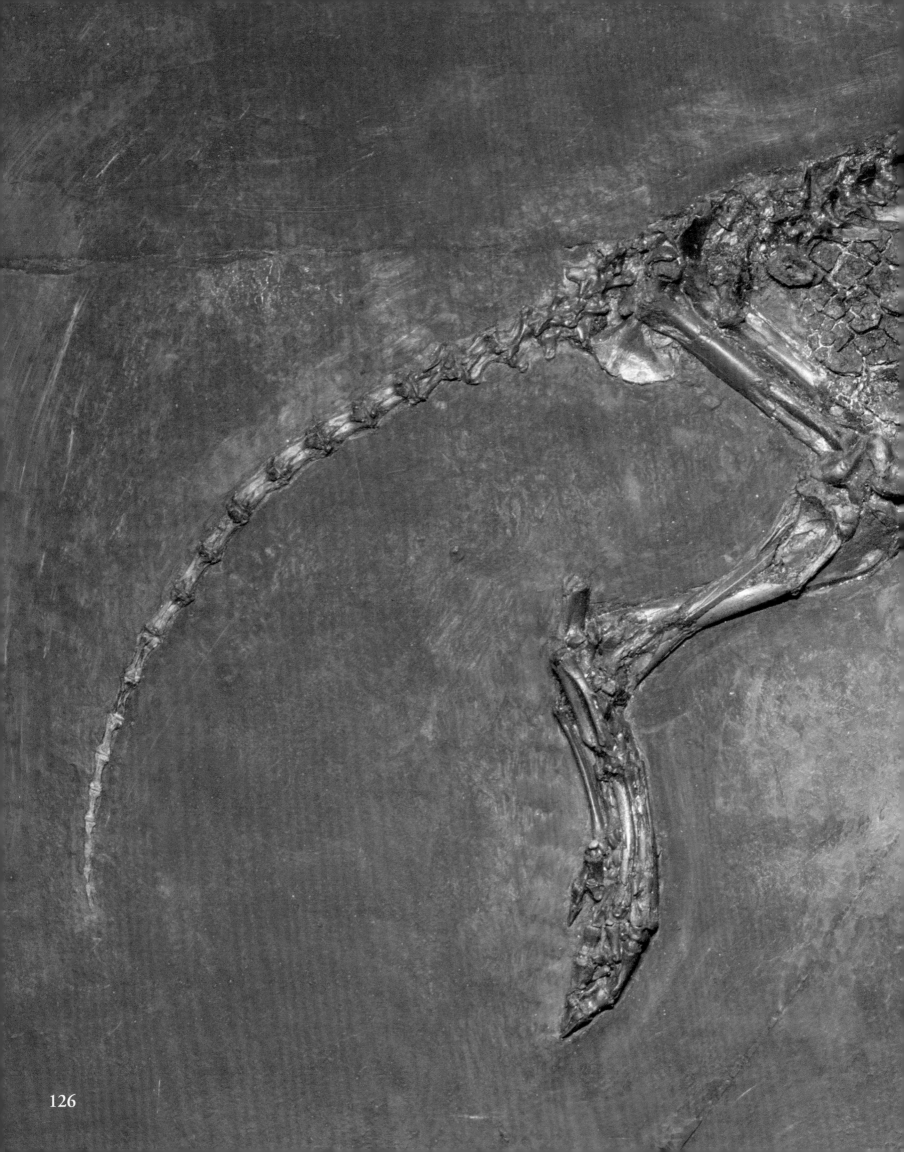

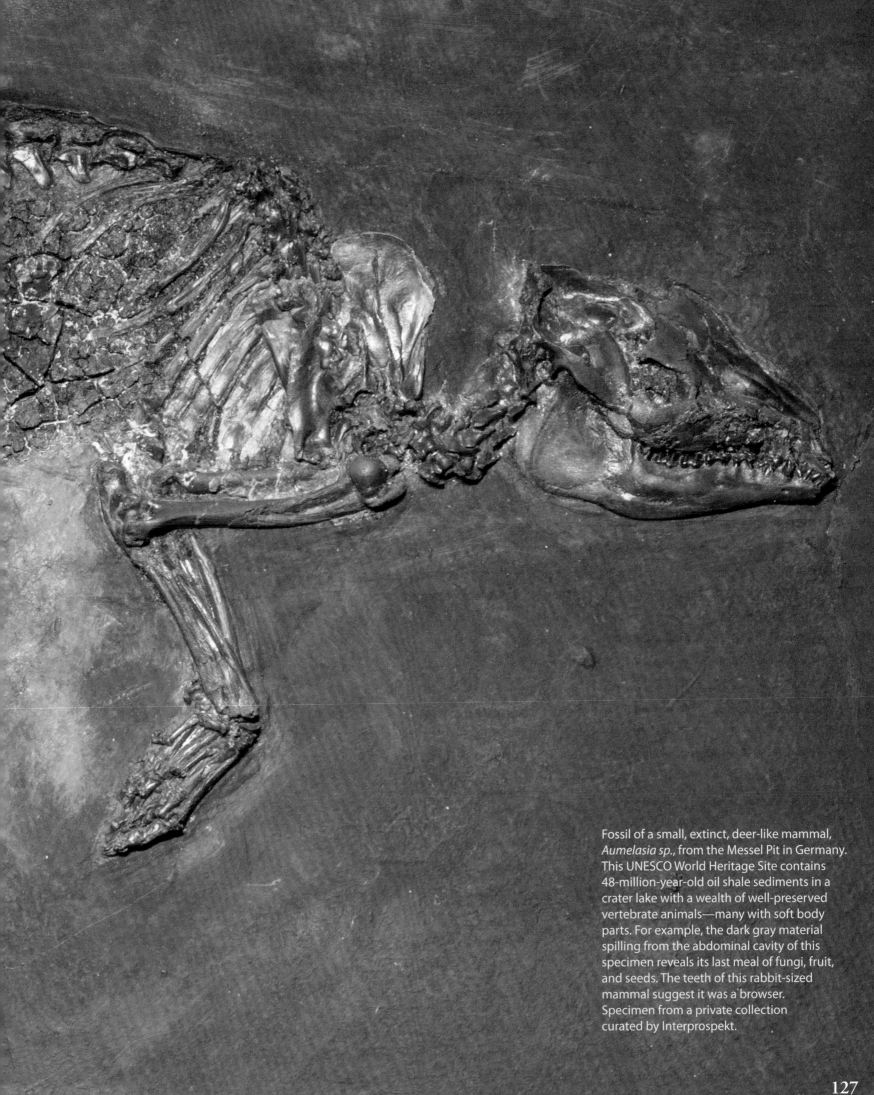

Fossil of a small, extinct, deer-like mammal, *Aumelasia sp.*, from the Messel Pit in Germany. This UNESCO World Heritage Site contains 48-million-year-old oil shale sediments in a crater lake with a wealth of well-preserved vertebrate animals—many with soft body parts. For example, the dark gray material spilling from the abdominal cavity of this specimen reveals its last meal of fungi, fruit, and seeds. The teeth of this rabbit-sized mammal suggest it was a browser. Specimen from a private collection curated by Interprospekt.

"A strictly terrestrial animal, by occasionally hunting for food in shallow water, then in streams or lakes, might at last be converted into an animal so thoroughly aquatic as to brave the open ocean."

~ Charles Darwin, from *The Origin of Species*, 1859

EARLY WHALE SKELETON

Cetaceans—whales and dolphins—evolved from terrestrial animals, and the most primitive whales had well developed pelvic bones and rear legs. Over time, as whales became more sea-worthy, bones of the pelvis shrank in size and lost their attachment to the spine. Today's cetaceans still have these bones—they are inconspicuous and embedded in muscle tissue. As you can see in the skeleton of this early whale—*Cynthiacetus* from Morocco—its pelvis and rear limbs were small but clearly formed. At this point in whale evolution (about 40–34 mya), the pelvis was no longer attached to the spine. Specimen courtesy of Serge Xerri of Rabat, Morocco.

In the ocean, the Oligocene was the golden age for **cetaceans**—whales, porpoises, and dolphins. Whales are distantly related to the hippopotamus and got their start about 50 mya—on the shore, not in the sea. Fossils of one of their Eocene wolf-sized ancestors, *Pakicetus*, had four legs for walking on land and an elongated skull with a bony wall around the middle ear, a feature found only in whales. It was evidently a shore-dweller that ate fish. Fossils of fully aquatic relatives clearly show the gradual transition from legs to flippers. The Oligocene whale *Mammalodon* was another evolutionary intermediate—it possessed both teeth and bristly plates of baleen for filter-feeding. In contrast, living whales fall into two major categories, those with teeth and those with baleen—none have a combination of both. The blue whale is a baleen variety with the distinction of being the biggest creature to have lived on Earth. Its heart is the size of a small car!

When and how early land mammals populated the Southern Hemisphere has historically received relatively little attention from the paleontological community, and overall, fossils have been scarce. There are more questions and opinions than solid answers. Keep in mind that failure to find fossils doesn't mean the animals were not present. Carcasses don't always get preserved as fossils; and fossils that are buried may never be found, especially in forested environments.

As the worldwide drying and cooling trend continued beyond the Oligocene—into the **Miocene** and **Pliocene** epochs—the parade of grazing animals in the Americas was joined by woolly mammoths, fearsome saber-tooth cats, giant ground sloths, and armadillos the size of a classic Volkswagen Beetle. The sloth, a South American native, evidently reached North America by "island hopping" before the **Central American land bridge** opened for two-way traffic in the Pliocene. Most fauna moved from north to south, displacing much of South America's mammalian fauna. This also marked the end of the southern reign of "terror birds," although one, *Titanus*, migrated north and succeeded in reaching southeastern parts of the USA. Australia had an even stranger assortment of wildlife—giant kangaroos and wombats, marsupial lions, monstrous monitor lizards, horned tortoises, and *Bullockornis*, better known as "demon ducks." Extinct ancestors of humans arose and diversified in the Old World, perhaps multiple times.

Our family tree is a patchy one that is frequently revised as new fossils are found. It is clearly more complex than we once thought. To the best of our knowledge, true primates first appeared early in the Cenozoic Era about 66–56 mya. This group gave rise to **prosimians**—a primate group including lemurs and tarsiers—and **anthropoids**, higher primates that led to monkeys, modern apes, and eventually humans. Monkey ancestors

GLYPTODONT

Armadillos and anteaters are closely related to sloths, and all three groups originated in South America. Automobile-sized armadillos—the **glyptodonts**—spread north into southern parts of North America. These grazing herbivores were heavily armored with a rigid shell of bony plates in their skin. Like sloths and anteaters, they vanished from the fossil record about 10,000 years ago near the end of the last Ice Age (= Pleistocene). The glyptodont shown here, *Panochthus tuberculatus*, is from the Pampas Beds of southern Uruguay, a skeletal cast courtesy of Gaston Design in Colorado, USA.

Living armadillos are much, much smaller—the fairy armadillo is no bigger than a chipmunk; and an average "giant" armadillo weighs about 60 pounds (27 kg), with a total length of 35 inches (89 cm), from its nose to back of its shell. The familiar nine-banded armadillo is the only species with a range from South to North America. And like all living armadillos, its shell is flexible.

somehow found their way to South America about 40–50 mya, where they evolved into a distinctive group with widely separated nostrils and long prehensile tails that could grasp branches. And at about the same time, Old World monkeys started their own evolutionary track to becoming modern baboons and macaques, while human ancestors followed another path. A "dawn ape" appeared about 30 mya, and the great apes *Dryopithecus* and *Ramapithecus* surfaced 15–25 mya in Africa and Eurasia. *Ramapithecus* might have given rise to *Australopithecus*, believed to be the lineage from which humans evolved 2–3 mya.

The **species group *Homo***, which includes Neanderthal Man, Cro-Magnon Man, and modern man (*Homo sapiens*), is still being sorted out. By the end of the **Pleistocene Epoch** (2.58 mya–11,700 years ago), *Homo sapiens* had spread to nearly every part of our planet. The Pleistocene was a time of repeated glaciations, often referred to as the last Ice Age. Many species perished during this time. Among them were the famous mega-mammals—mammoths, woolly rhinos, giant ground sloths, saber-tooth cats, and cave bears, for example. On the other hand, many familiar mammals survived, including apes, cattle, deer, rodents, kangaroos, wolves, and more. Evolutionary biologists and anthropologists continue to debate the role our ancestors played in at least some of these Pleistocene extinctions.

GIANT GROUND SLOTH

Sixty different kinds of giant ground sloths have been discovered. These large, lumbering vegetarians evolved in South America about 35 mya and began migrating into North America about 8 mya. They varied in size from that of a bear to an elephant. In the American West, Shasta ground sloths sheltered in dry caves, where deep deposits of preserved sloth dung have allowed biologists to accurately reconstruct their paleo-environment and diet. Ground sloth extinctions followed the spread of humans in the Americas late in the Pleistocene epoch, about 16,000–11,000 years ago, but the role played by mankind is still being debated. This painting of the Panamerican giant ground sloth, *Eremotherium laurillardi* (a species found in Florida) is part of an exhibit in the Museum of Ancient Life/ Thanksgiving Point Institute, Lehi, Utah, USA.

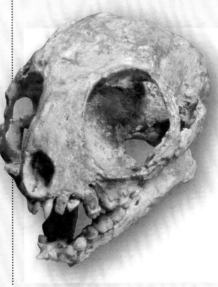

EARLY PRIMATE

Golf-ball-size skull of an early primate, *Necrolemur antiquus*, an ancestor of living tarsiers— small big-eyed primates found only in Southeast Asia. This specimen of Middle to Late Eocene age (about 38 mya) is from phosphate deposits in the Quercy Region of France. Specimen courtesy of Charlie Magovern.

EVOLUTION OF PLANTS

For nearly half of Earth's 4.6-billion-year history, there was virtually no oxygen in the atmosphere. This started to change about 3.5 bya when the earliest known life forms, cyanobacteria, began populating the ocean. Equipped with the green pigment chlorophyll, these marine microbes could harness the sun's energy to manufacture food from water and carbon dioxide—photosynthesis—releasing oxygen gas as a byproduct. Oxygen that accumulated in the sea began slowly escaping into the atmosphere. Geoscientists believe that for the next billion years the atmosphere remained a mixture of mostly nitrogen, carbon dioxide, and methane, with oxygen as a minor component at 2-4%. Today, about 21% of the air we breathe is oxygen and most of the rest is nitrogen (78%).

So how do geochemists estimate atmospheric oxygen levels in deep time? They examine rocks containing chemical traces of molecules that form only in the presence of oxygen. These rocks are mostly volcanic in origin, and the greater the concentration of these molecules, the more oxygen must have been present. Or conversely, finding molecules that form only in the absence of oxygen indicates that no oxygen was in the atmosphere at that place and point in time.

Some 2.3 bya, levels of free oxygen rose dramatically in Earth's atmosphere, known as the **Great Oxidation Event**. Scientists agree that it happened, but the cause or causes have long remained a mystery. Perhaps the change came when cyanobacteria diversified, proliferated, and formed aggregations. Rising oxygen levels may have attacked rocks chemically, releasing minerals that further nourished and accelerated microbial blooms, producing even more oxygen. And cyanobacteria may have played a significant role in early plant evolution. Many

Bright green algae like the one shown here—"gutweed" (= *Ulva intestinalis*)—thrive in cool seawater along rocky coasts of the world. "Sea lettuce" (*Ulva lactuca*), gutweed, and other species of *Ulva* are edible, nutritious, and commonly added to soups and salads in Scandinavia, the UK, China, and Japan. These algae are packed with protein, complex carbohydrates, soluble fiber, vitamins, and minerals, especially iron. But beware! Plants growing in water contaminated with toxic heavy metals can be dangerous to eat. Photographed on Appledore Island, Isles of Shoals, Maine, USA.

living species can convert nitrogen gas into organic compounds needed for plant growth. What we know for certain is that rising levels of oxygen provided a source of energy that triggered the evolution of complex animal life, which blossomed during the Cambrian 541–485 mya.

The evolution of land plants, which probably began about 700–800 mya (according to genetic studies), must have contributed dramatically to the rise in atmospheric oxygen. About 480 mya, a spike in atmospheric oxygen appeared in the geological record along with fossils of the earliest known land plants—mosses, liverworts, and hornworts. Since then Earth's oxygen budget has experienced many boom and bust cycles through time, heavily influenced by geophysical forces. Volcanic outgassing, for example, can generate new compounds that suck up oxygen faster than plants produce it. And when continents drifted into cooler, drier locations, forests shrank and so did their oxygen output. Surface rocks rich in iron also use oxygen as they rust (oxidize), and microbes that decompose dead organisms consume oxygen too. Making sense of complex phenomena in deep time requires an open mind. The best explanations usually come from multiple, overlapping lines of inquiry.

PRAISE FOR ALGAE

Algae (plural for *alga*) are much more than "pond scum." Most of the bad press given to algae can be traced to problems of our own making (see Chapter 7, *Extinction Events*). Algae come in a vast array of sizes, colors, and shapes—from free-floating microscopic forms to red feathery clumps on the sea floor and luxurious forests of giant kelp. Algae not only enrich our atmosphere with oxygen, they are the foundation for most of our planet's food chains/webs. They have given us pharmaceutical compounds used in our fight against bacteria, viruses, and cancers. Much of the oil we extract comes from Cretaceous deposits of marine algae, and some oily micro-algae are mighty giants on the biofuel frontier. They can make oil faster than oilseed crops like soybeans and sunflowers. And unlike traditional crops that require fertile farmland, microalgae can be grown in water anywhere, even on rooftops! Shown here is a research lab at the Arizona Center for Algae Technology and Innovation at Arizona State University.

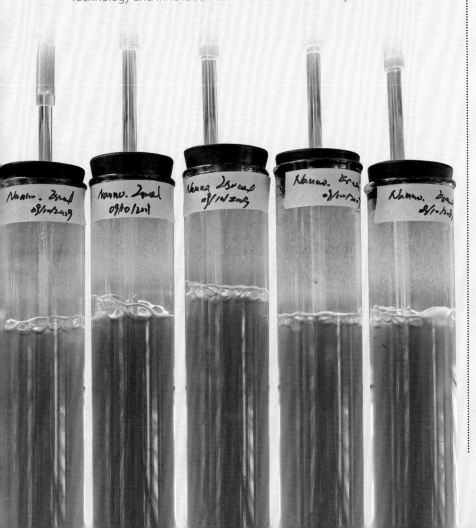

ALIENS ON THE HOME FRONT

Slime "molds" are slippery creatures that defy clean classification. Like algae, they aren't considered plants, fungi, or animals—they have been placed in a "wastebasket" kingdom, **Protista**. Slime molds slither around, amoeba-like, in damp places. Some remain single-celled with one nucleus; others take the form of a "super-cell" with thousands of nuclei (a *plasmodium*) that may cover an area the size of a bath towel. They actively hunt and engulf bacteria (some unicellular algae also eat bacteria!). And when conditions are right, they shape-shift into tiny mushroom-like fruiting bodies filled with spores (*sporangia*). The spores can survive periods of drought for years. When rain comes and the spores hatch, these single-celled siblings use chemical cues to find each other. They then merge to form a blob-like plasmodium. Lab experiments with slime molds have shown that a plasmodium can even navigate its way through a maze to find food and learn from experiences.

The soft bodies of slime molds don't preserve well, so it's no surprise that their fossil record is nearly empty—only a few have been found in amber. But genetic studies suggest that ancestral slime molds evolved at least 600 mya.

Slime molds shown here—bright pink plasmodium blobs of *Lycogala epidendrum* (**above**); and orange sporangia of *Trichia sp.* (**below**)—were photographed in the Chiricahua Mountains, Arizona.

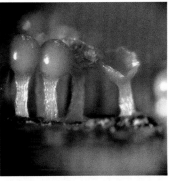

Land plants that cover much of Earth's surface began their rapid evolution and radiation about 450 mya, stemming from an ancestral group of simple freshwater algae called **charophytes**, which remain abundant today. These are true plants with **chloroplasts**, specialized chlorophyll containers (*organelles*) where photosynthesis takes place within their cells. Although sometimes called blue-green "algae," cyanobacteria differ from these green algae and other true plants in many ways. Most noticeably, cyanobacteria lack nuclei and other membrane-bound structures like mitochondria and chloroplasts.

Lichens, those colorful pioneers that often look like paint spattered on rocks in dry places, are **fungi** (plural of *fungus*) that have teamed up with cyanobacteria and green algae. They live together and benefit one another in a *symbiotic* relationship. The fungus is sponge-like and contributes shelter, moisture, and minerals to its green, photosynthetic, food-sharing partners. Molecular studies of modern species suggest that lichens were among the first organisms to colonize land, possibly dating to Precambrian times (more than 541 mya).

Fungi are not plants—they belong to Kingdom Fungi, which includes mushrooms, yeast, and molds. Their spongy tissues do not preserve well, and most clues to their existence are microscopic and difficult to interpret. Although still inconclusive, chemical evidence is mounting that fungi might have begun moving onto land more than 700 mya. Ninety-million-year-old mushrooms in amber are among the few undisputed fungus fossils visible to the unaided eye.

Dead man's fingers, a fungus growing from a moss-covered log in the Amazon Basin of Ecuador. Both fungi and mosses reproduce with dust-like spores, comparable to "seeds". Look closely at the moss—rising above it you can see thread-like stalks tipped with tiny capsules (= sporangia) that contain spores. Many fungi—dead man's fingers and mushrooms, for example—produce spores in a fleshy fruiting body that's good to eat, or not!

Spring-fed Proxy Falls cascades over moss-covered basalt columns in Willamette National Forest, OR, USA.

Lichens are quick to colonize raw rock, even in deserts, as seen on this dry volcanic terrain in Utah's Snow Canyon State Park, USA. Given the abundance and diversity of lichens on rocks, on trees, and in biological soil crust, it's astonishing that before 2017, only 15 lichen fossils had been reported, and nine of these were found in Baltic amber. Then in 2017, more than 150 new specimens were discovered in the same amber deposit from 47–24 mya. Many lichens live in places that don't favor preservation as fossils, and they are easy to overlook. So far, the oldest confirmed specimen is from Rhynie chert of Early Devonian age (412–400 mya) in Scotland.

In recent lab experiments, Zhi-Yan Du and his colleagues at Michigan State University reported that a modern soil fungus and a marine alga can quickly form a symbiotic partnership. Within six days, the two had made physical contact and began exchanging nutrients. Within two months, some of the algal cells had moved inside the threadlike filaments (*hyphae*) of the fungus. The photosynthetic algal cells continued to function, grow, and divide within the fungal cells—this is a biological first that has not been seen in the wild. But it's not hard to imagine that similar relationships between algae and fungi could have existed more than a billion years ago.

Through the ages, land plants and animals have been inseparable. And right from the start, plant anatomy has been shaped by gravity. Mosses, liverworts, and other pioneering plants lacked specialized tissues for transporting water and nutrients. As a result, these little **non-vascular plants** hugged moist ground, and their root-like *rhizoids* gave them support. Like today's ferns and other primitive plants, they released reproductive cells called **spores**. These cells have the ability to grow into new individuals without fertilization by another sex cell (*gamete*). Spore capsules (*sporangia*) and spores are typically the only remains of these plants found as fossils, alongside early land animals—insects, spiders, millipedes, and the like.

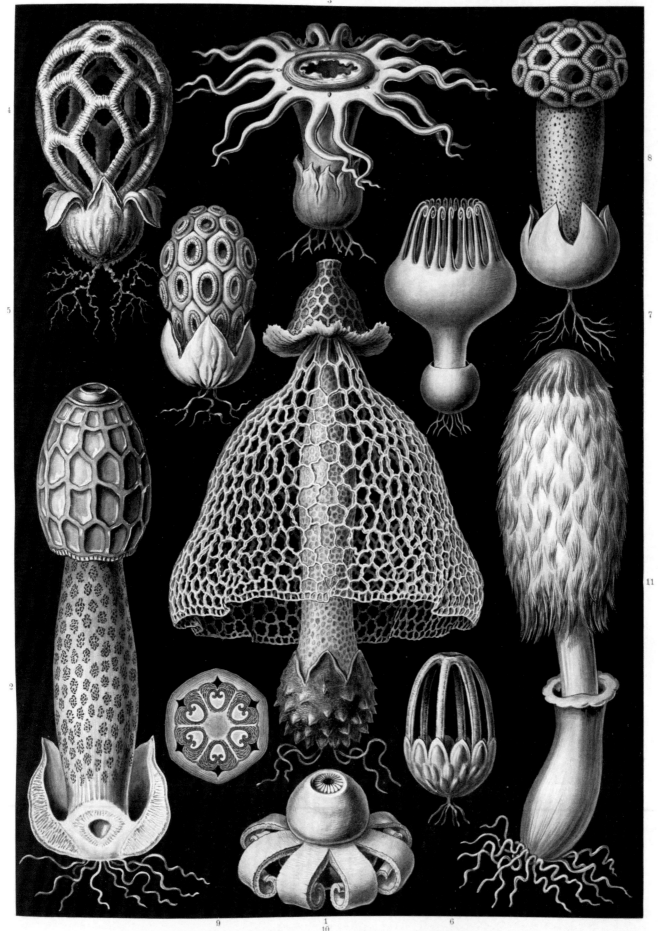

Basimycetes. — Schwammpilze.

ORGANISMS IN THE KINGDOM FUNGI ARE NEITHER PLANTS NOR ANIMALS—
UNLIKE PLANTS, THEY CANNOT MANUFACTURE FOOD.

Hepaticae. — Lebermoose.

LIVERWORTS—BELIEVED TO BE AMONG THE FIRST PLANTS TO COLONIZE LAND

Vascular plants can better defy gravity by being strengthened with internal vessels for moving fluids. *Cooksonia*, one of the earliest vascular plants in the fossil record, was only ankle-high and had naked, branching, upright stems with sporangia on the tips. It was named after paleobotanist Isabel Cookson (1893–1973), one of Australia's first professional women scientists. Since its discovery, this important group of extinct primitive plants has been found in mid-Silurian to early Devonian deposits throughout the Eastern and Western Hemispheres, dated to 433–393 mya.

Near the end of the Devonian Period, vascular plants began to increase in size and abundance. They towered above their predecessors, thanks to a new chemical innovation, **lignin**, a polymer that allows the formation of wood and bark. Combining the power of lignin with **cellulose**—a tough compound in the cell walls of all plants—trees have been able to send roots deep into the ground and attain giant stature that dwarfs the largest animals to have set foot on Earth. These new plants also developed complex leaves for harvesting sunlight.

Ferns, clubmosses, and horsetails are among the first plants with a tough protective covering (= *cuticle*) and vascular tissues. They have sturdy upright stems and creeping horizontal stems (= *rhizomes*) with true roots, structures evident in this 92 million-year-old fern fossil from Brazil's Santana Formation (**above**)—specimen courtesy of Annesuse Raquet-Schwickert.

Like their extinct ancestors, living plants in these three primitive groups all depend upon spores for reproduction. Ferns produce spores in circular patches, *sori*, on the underside of their fronds—see backlit example **below**, photographed in the Hawaiian Islands.

Clubmosses and some horsetails package their spores in specialized cone-like structures called *strobili*, as seen here in this living horsetail (**left**) photographed in Canyonlands National Park, Utah, USA. Its fossilized counter-part is from Wyoming's Green River Formation—specimen courtesy of the Black Hills Institute of Geological Research. The living clubmoss in the **large circle below**, *Lycopodium sp.*, was photographed at Freshwater Lake on the Caribbean island of Dominica—no cones shown.

FOSSILS INSIDE OUT

Ferns, clubmosses, horsetails, and scale trees dominated great swamp forests that spread during the Carboniferous Period (359–299 mya), forests that were slowly transformed into our planet's rich coal deposits. Most of these plants had cone-like reproductive organs filled with large spores. Needless to say, we have no shortage of impressive plant fossils from this period. And there were a few *big* invertebrate animals too, such as dragonflies with wingspans up to 28 inches (70 cm) and giant millipedes the length of a giraffe's leg!

In the plant world, the evolution of the seed was as important to terrestrial life as the innovation of the amniotic egg was to reptiles. The first seed-producing plants—**pteridosperms**, commonly called **seed "ferns"**—had fernlike leaves and some were tree-like. But they are unrelated to modern tree ferns or any other true ferns. Seed ferns proliferated in Late Carboniferous swamp forests, but then disappeared from the fossil record. One lineage of this diverse group appears to be ancestral to cycads. Conifers, ginkgos, and cycads—**gymnosperms**—evidently displaced the seed ferns.

Bark of a **scale tree**, *Lepidodendron* (**below**), showing its typical pattern of leaf scars. Like clubmosses and quillworts, this primitive genus of plants reproduced by spores in "cones," not by seed. Scale trees were the first large land plants—some taller than the tallest sauropod dinosaurs. But they did not live among dinosaurs—they were abundant in coal-forming Carboniferous forests and disappeared before the Age of Reptiles. This specimen from Larksville, Pennsylvania is on display at the Wyoming Dinosaur Center.

Left: German etching of Carboniferous swamp forest with scale tree in front/left; from Meyers Konversationslexikon (1885-1890)/ Bibliographisches Institute.

Seed fern fronds, most likely *Pecopteris*, from Late Carboniferous deposits in Leon Province, northwestern Spain. Specimen courtesy of Chris Moore Fossils.

Modern-looking conifers—pines, yews, redwoods, cypress, and monkey-puzzle trees (*Araucaria*)—were well-established early in the Mesozoic Era (252–66 mya). Gymnosperms have no flowers or fruit, and their seeds develop in female cones, where they are fertilized by wind-blown pollen from small male cones. The female cones are packed with protein, and cycad seeds have been found in fossilized stomach contents of an Argentinean hadrosaur, *Isaberrysaura*, clear evidence that some dinosaurs ate cycad cones.

Botanists know that many parts of living cycads are loaded with toxins, so if early cycads had these chemicals, dinosaurs either avoided the toxic tissues or simply ate them. Keep in mind that foods poisonous to mammals can be a reptile's favorite. In fact, modern Caribbean rock iguanas and Galápagos tortoises love toxic manchineel fruit and swallow them whole. Dubbed "the little apple of death" by conquistadors, this sweet-smelling, crabapple-sized fruit is loaded with toxins, as are all other parts of this plant.

Fossilized wood of an ancient conifer, *Woodworthia arizonica* (**below**), showing exterior pits on the surface, which might have been the location of spines or small undeveloped branches. No fossil leaves, cones, or seeds have been found, so this primitive tree has not yet been assigned to a plant family. Specimen is of Late Triassic age, from the Chinle Formation in Petrified Forest National Park, Arizona—it resides in the park's museum collection.

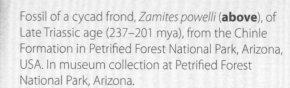

Fossil of a cycad frond, *Zamites powelli* (**above**), of Late Triassic age (237–201 mya), from the Chinle Formation in Petrified Forest National Park, Arizona, USA. In museum collection at Petrified Forest National Park, Arizona.

Although the cycad lineage itself is ancient—going back in the fossil record at least 280 million years—most living species have evolved in the last 12 million years. This sago cycad, *Cycas revoluta*, with a mature male cone (**inset**), is one of about 250 modern species. Native to southern Japan, it's commonly used as an ornamental landscape plant. Grown by Barbara Rogers in Tucson, Arizona.

ARAUCARIA TREES

A spectacular group of evergreen trees, family Araucariaceae, dates back to the Age of Reptiles. Fossils have been found in India, Australia, England, Europe, Africa, and North America, including the giant mineralized logs in Arizona's Petrified Forest National Park. Only 20 species remain today, all native to South America and islands of the South Pacific, and many are grown as ornamental trees. The Norfolk Island "pine" (*Araucaria heterophylla*) is perhaps the most familiar, often sold as potted plants at Christmastime. It has spiraling scale-like leaves, similar to those seen on young branches of this hoop "pine," *A. cunninghamii* (**far right**). Others, like the monkey-puzzle tree, *A. araucana*, have stiff, dagger-like foliage that no monkey would want to touch!

Whole and cut petrified *Araucaria mirabilis* cones (**above**) from Cerro Cuadrado region of Patagonia, Argentina. Specimen courtesy of Larry & Pat Martin.

Top: Illustration by Paul Mirocha.

DAWN REDWOOD

Conifer trees in the genus *Metasequoia* were fairly common in North America and Asia late in the Mesozoic and early Cenozoic eras. The fossil shown here (**left**)—courtesy of Mark and Karen Haverstein/Lowcountry Geologic—is of Oligocene age from the Muddy Creek Formation in southwestern Montana.

In 1941, a single living population of these primitive trees—named the **dawn redwood**, *Metasequoia glyptostroboides*—was discovered in a remote mountain valley in China. Today, threatened by deforestation, only about 10,000 survive in the wild. The author photographed this tree (**inset at left**)—grown from seed—at the Missouri Botanical Garden in St. Louis, MO, USA.

Flowering plants, the **angiosperms**, are the most diverse and successful members of the plant kingdom today. They are found in virtually every habitat on Earth and come in every conceivable shape, ranging from vines, succulents, and grasses to woody shrubs and hardwood trees. Their success, which began back in the Mesozoic Era at least 130 mya, has been driven by symbiotic relationships with their animal pollinators, especially insects, birds, and bats, and perhaps even dinosaurs. From a flower's point of view, attracting an animal to move pollen to the right spot for fertilization and seed growth is much more reliable than letting wind do the job.

This complex evolutionary dance between flowers and their pollinators has created sophisticated partnerships (**co-evolution**), loaded with floral trickery. Most flowers reward pollinators with a sweet nectar treat for their services. But some just look and smell like the insect they attract—when a suitor attempts to mate with this alluring floral fake, all it gets from the deal is a coat of pollen. And some plants use irresistible aromas to guide bees to a hair-like trigger that fires a sticky pack of pollen onto the visitor's back. Other flowers are gatekeepers, trapping insects just long enough to be certain they are well covered with pollen before setting them free. Still others have created deceptive floral parts that look like fruit to entice foraging birds—one tug on the hollow "fruit" showers the animal's head in pollen.

Such innovations seem endless, all in the name of moving pollen from flower to flower for successful sexual reproduction. Flowering plants represent 80% of all known species in the plant kingdom, and insects—their most common pollinators—are the most diverse and abundant multicellular organisms on Earth. These are the evolutionary winners of the Cenozoic Era!

We love our flowering plants. They feed us, shelter us, provide new medicines, and bring us joy. They add oxygen to the atmosphere, cool our planet, and trap carbon dioxide that drives global warming. The world's great rainforests, like those in the Amazon Basin, are essential storehouses of plant and animal diversity, but are they the "lungs of our planet"? They are not. Botanists have concluded that although land plants release a lot of oxygen into the atmosphere, the microbes and fungi that decompose forest leaf litter and detritus consume about the same amount of oxygen. So a forest's net oxygen output is close to zero. Yet without trees, human civilization and the very fabric of life on Earth would quickly unravel.

FOUL-SMELLING FLOWERS

Carrion- or dung-loving flies and beetles find stinky flowers irresistible. These insects seek decaying organic matter for food or a place to lay their eggs. Consider, for example, South African starfish flowers (*Stapelia gigantea*) **shown below**—they are large, hairy, and smell like rotting meat. Flies attracted to the center of the flower snag pollen-packs on their legs or mouthparts. After being carried to another flower, the pollen bags snap off in the right place—if the flower is lucky—allowing fertilization and seed development. The flower's disguise is so convincing, flies often lay eggs on their hairy petals. The larvae hatch there, find no food, and die—collateral damage. The flowers do not trap and eat flies, and they do reward their pollinators with nectar that is super-rich in amino acids. The author grew and photographed these flowers in Tucson, Arizona.

FOSSILS INSIDE OUT

POLLINATION PUZZLER

Famous British explorer, botanical illustrator, and activist Margaret Mee (1909-1988) was obsessed with finding the elusive moonflower cactus in bloom (*Selenicereus wittii*, **shown here**). This rare plant lives in flooded forests of the Amazon Basin, anchoring itself to a tree trunk with roots that grow from flattened stems. Its fragrant white flowers open at night—for only a few hours—during the height of flood season. For a glimpse of the story behind this intrepid woman—and her 20-year search for a flower to paint in the wild—watch the 2012 Brazilian documentary *Margaret Mee and the Moonflower*.

To reach nectar at the bottom of the long floral tube of this cactus, a pollinator would need a very long tongue. And indeed, there are flying insects with a 10-inch (25-cm) coiled proboscis for probing moonflowers—hawkmoths. Based on his theory of natural selection, Charles Darwin had predicted the existence of such moths while studying the anatomy of the Madagascar star orchid in 1862. This orchid's nectary lies hidden at the bottom of a long floral spur. A candidate moth was discovered 20 years later, the year Darwin died, but no one had actually observed this insect pollinating the orchid until 1992. This moonflower specimen, grown by Mark Dimmitt, was photographed in Tucson by the author.

INSECT FOSSILS

Early in the Cretaceous, about 108 mya, flowering plants were beginning to diversify along with their insect pollinators. Fossilized insects from Brazil's Crato Formation provide an important window to their evolution, exemplified by this wasp-like insect and a beautifully preserved butterfly or moth wing. Specimens courtesy of Merv Feick/Indiana9 Fossils.

Nepenthaceae. — Kannenpflanzen.

PITCHER PLANTS—FLOWERING PLANTS THAT CAN TRAP INSECTS OR
CATCH FALLING LEAVES NEEDED FOR NITROGEN IN NUTRIENT-POOR SOILS

Tineida. — Motten.

FUNGUS MOTHS (FAMILY TINEIDAE)—THEIR LARVAE ARE UNUSUAL;
MOST FEED ON FUNGI, LICHENS, OR DETRITUS INSTEAD OF LIVING PLANTS

EXTINCTION EVENTS

7

Our planet has had a long history of winners and losers, mostly losers in the context of geological time. Paleobiologists believe that over 90% of all life forms to have existed on Earth are now gone.

By studying the composition of rocks, sediments, and the fossil record, scientists have identified five major mass extinction events (see *History of Life* chart in Chapter 4) and plenty of minor ones.

At the end of the Age of Reptiles about 66 million years ago, a gigantic asteroid smacked into Earth near the coast of Mexico's Yucatan Peninsula. This is perhaps the clearest example of cause-and-effect leading to a mass extinction. Studies of the asteroid's submarine impact crater and rocks of the same age elsewhere in the world continue to support this theory. Further evidence suggests that widespread volcanic eruptions in west-central India set the stage for this extinction event before the collision.

ROCKS FROM SPACE

Barringer Meteor Crater is 4,150 feet (1265 m) across and 570 feet (174 m) deep. Meteorologists calculate that a 63,000-ton (about 57-million-kg) nickel-iron rock—a *meteorite*—the size of a small house crashed to Earth about 50,000 years ago, blasting this enormous hole in the northern Arizona desert. But this crater is dwarfed by Mexico's Chicxulub sub-sea asteroid crater of 66 mya.

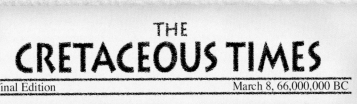

THE CRETACEOUS TIMES

Final Edition   March 8, 66,000,000 BC

DOOMSDAY ASTEROID COLLIDES WITH EARTH

By Mari N. Jensen

Chicxulub, Mexico (66 million years BC)

A fiery 6-mile-wide asteroid with a glowing tail plummeted into the continental shelf by Mexico's Yucatan Peninsula, vaporizing rocks and water and making a crater 112 miles wide. The scientific community predicts cataclysmic consequences for life on earth.

Globules of molten rock and clouds of sulfur dust are rising high into the atmosphere. As the tiny superheated pellets of rock fall back to earth like hot hail stones, forests around the world are bursting into flames, creating a death-dealing firestorm.

The whole world may catch fire, predict scientists.

And, although rain may fall, the sulfur dust kicked up by the asteroid's impact will turn that rain into a shower of sulfuric acid. As the acid rain drains into rivers and oceans, the change in water chemistry will kill many aquatic plants and animals.

Clouds of smoke and ash are boiling into the skies from the forest fires, making it impossible to see the sun. Meteorologists say the thick cloud layers of ash and sulfur dust may darken the skies for years. Those dark clouds may block the sun's rays so completely that plants that survive the fires will still die from lack of sunlight.

If that happens, life as we know it in the Cretaceous period would come to an abrupt end. Biologists predict most animals, from the majestic to the miniature, including the dinosaurs, will perish.

Rodents Pulling Your Tail?

An "eye-witness" report of a cosmic collision between an asteroid and Earth some 66 mya, based on current scientific evidence and theory. Artwork by David Fischer.

Episodes of intense freezing have also played an important role in Earth's early history. By the end of the Ordovician Period about 445 mya, most of the land masses present at that time had united into one supercontinent, Gondwana, with crustal plate collisions that built mountains. By then, Gondwana had drifted to the South Pole, and catastrophic drops in temperature brought on glaciation. As ice sheets formed, sea levels fell and shallow marine basins dried up. Most paleoecologists agree that these changes caused Earth's first mass extinction. Ordovician life forms were largely marine.

Causes of the other three mass extinctions have been more difficult to assess. Quite likely, multiple geophysical changes contributed to each cataclysm, triggered by continental drift and tectonic activity—spawning earthquakes, volcanoes, mountain-building, and the formation of ocean trenches. Whenever our planet rearranges its physical anatomy, everything else changes as well—its atmosphere, ocean currents, water chemistry, sea level, and climate. So too, living organisms can transform their physical environment. For example, dramatic shifts in prehistoric flora (including algae and bacteria) have altered the chemistry of Earth's atmosphere and aquatic environments. And the rise of terrestrial plant life pumped more oxygen into the atmosphere, opening the door for big animals to colonize land.

Our planet remains restless, allowing humans to experience these processes in action, on a smaller scale:

Exploding Lakes, a rarity today, might have been a common phenomenon in deep time when volcanoes were especially active. On 21 August 1986, Lake Nyos in the African Republic of Cameroon literally exploded at night . . . but without making a sound. Released from the depths of this submerged volcanic caldera, an enormous burst of carbon dioxide gas (CO_2) raced swiftly into nearby villages, silently suffocating 1,746 people (99% of the villagers) and 3,500 farm animals in minutes. Being heavier than air, CO_2 hugged the ground and rolled down valleys like an invisible flash flood while villagers slept. Magma beneath this deep lake leaks CO_2 into its water, and over hundreds of years the lake had turned into a pressurized time bomb. What triggered its sudden release remains a mystery.

Creeping Dead Zones form in the depths of large lakes and along the coasts of major continents. In these places, water near the bottom can become rich in CO_2 and depleted of oxygen (anoxic). The fossil record is full of *mass-mortality beds* where aquatic creatures were apparently starved of oxygen. Today, dead zones have been on the rise since 1970, especially where industrial and agricultural run-off adds nutrients to the water. As of 2008, more than 400 had been identified. The largest is in the Arabian Sea, which is deep, has little water movement, and becomes naturally stagnant. Rising temperatures from climate change will drive the expansion of dead zones worldwide.

MASS-MORTALITY PLATES

Massive die-offs of aquatic animals frequently appear in the fossil record—presumably triggered by oxygen depletion, drought, disease, temperature stress, food shortage, or an influx of toxic chemicals/micro-organisms. Two mass-mortality plates are shown here—trilobites, *Xenasaphus devexus*, from Middle Ordovician deposits near St. Petersburg, Russia (specimen courtesy of Tom Lindgren/GeoDecor); and a plate of bony fish, *Semionotus ornatus*, of Late Triassic age from the Seefeld Formation, Salzburg, Austria (specimen courtesy of Martin Goerlich/Eurofossils).

Harmful algal blooms have been on the rise along Florida's extensive coastline, leading to massive die-offs of sea life. For example, in 2016 a coffee-colored "brown tide" algal bloom in Indian River Lagoon south of Cape Canaveral killed sheepshead fish (in most of this photo), mullet, croaker, puffer fish, catfish, flounder, spadefish, and horseshoe crabs, among others—photo © Malcolm Denemark/Florida Today. The plankton responsible for brown tides are tiny—each about the width of a strand of spider silk—and they multiply to catastrophic numbers in altered environments. Every HAB event in Florida has had a multi-million-dollar economic impact related to lost fisheries, lost wildlife, lost recreation and tourism, beach clean-up, and respiratory and digestive illnesses.

The Western Hemisphere's largest dead zone develops every spring in the northern part of the Gulf of Mexico. The cycle begins when farmers fertilize their fields in midwestern states of the USA. Rainwater transports fertilizer along with natural nutrients into the Mississippi River, eventually reaching the Gulf. This "nutrient pollution" nourishes an over-growth of microscopic phytoplankton, especially algae, producing an algal bloom. When this build-up begins to die and sinks to the sea floor, bacteria that decompose their remains consume oxygen and release CO_2. Anoxic water has seriously damaged commercial fisheries in the Gulf of Mexico, and the path to recovery has been elusive.

Red Tides also stem from an excess of nutrients that flow into coastal marine waters. Algal blooms are nothing new on our planet, but human activities have altered and intensified their effects in today's aquatic ecosystems. Algae produce a lot of oxygen and normally play a vital role in the food chain. But a few single-celled species of diatoms and dinoflagellates secrete defensive toxins, among the most potent in nature. And during their massive, often visible blooms, the ocean seems to be stained with red or brown dye, a red tide. These "harmful algal blooms" (called HABs by marine biologists) can take a heavy toll on sea creatures, including crabs, sea turtles, manatees, dolphins, and fish of all sizes, from eels to whale sharks. Even seabirds can die from poison passed up the food chain.

Tambora's Lesson: In 1815, on an Indonesian island just east of Bali, Mt. Tambora literally blew its top. It spewed out a stormy cloud of ash, pumice, and hot toxic gasses (a *pyroclastic flow*) that engulfed an entire peninsula, killing 12,000 people on contact. This was the deadliest known eruption on our planet in the past 10,000 years, ten times more powerful than the well-publicized Krakatoa eruption of 1883.

Tambora's explosion set into motion an observable, uncontrollable chain reaction of events that led to short-term global death and destruction. Another 80,000 people died in earthquake-triggered tsunamis and from infection, disease, or starvation. An airborne cloud of volcanic ash cast a dark shadow over Southeast Asia for a week. And millions of tons of sulfur dioxide blasted into the stratosphere began circling the globe. Sulfate ions (oxidized sulfur dioxide) reflected sunlight back into space. Temperatures dropped worldwide, and even though the drop averaged only about 1°F (0.6°C), the following year (1816) became known as the "year without a summer." Crops failed in North America, and Europe had its worst famine of the 19th century. Tambora offers a healthy reminder that we live in a delicately balanced environment, and seemingly small shifts in this balance can have far-reaching global consequences.

Atmospheric CO_2 and Methane: Despite rumors to the contrary, geologists now know that emissions from today's active volcanoes contribute little to climate change—less than 1% of the volume of greenhouse gases generated by humans. Carbon dioxide is our biggest enemy. Methane is second. Methane is far less abundant and has a shorter lifespan than carbon dioxide; but molecule-per-molecule, methane's global warming effect is significantly greater than that of carbon dioxide.

Most atmospheric methane comes from decomposition of organic matter in natural wetlands and rice paddies; from landfills and livestock waste; from forest fires and wood-burning; and from coal mines and leaks in oil and gas operations. Geoscientists have discovered that with global warming, melting glaciers and thawing ground in the Arctic are now releasing a lot of methane that was previously contained under the ice. Formerly frozen wetlands provide a methane-making cocktail of trapped organic matter, microbes, low oxygen, and water. Other researchers are studying bacterial strains that might prove useful as "biofilters" to control methane emissions.

Early man has always been a hunter with an impact on his environment—but to what extent? Scientists continue to debate the role of humans in the loss of Ice Age mammals that existed 5,000–50,000 years ago. Yet while re-tracing our footsteps through more recent time (over the last 5,000 years), anthropologists and paleobiologists have found one correlation that repeats itself: wildlife populations, especially those of giant land mammals, birds, and reptiles, have declined or have been pushed to the brink of extinction in the wake of human expansion.

Some remarkable animals, most of them island species that co-existed with the first waves of modern human settlers, have now been gone for more than a century. Their stories have been told through folklore, journal entries, paintings, and museum specimens (skins, bones, or fossils). Among those wiped out in the presence of humankind during the past 100–1,000

years are Madagascar's elephant bird and gorilla-sized sloth lemur, and New Zealand's great flightless moa and giant parrot "Squawkzilla." Other species that vanished are the dodo in Mauritius, the British great auk, the Formosan clouded leopard, Reunion Island giant tortoise, Cuban red macaw, Jamaican poorwill, Japanese sea lion, Steller's sea cow of Europe, Australia's toolache wallaby, and the North American passenger pigeon. Little doubt remains that overhunting and destruction of critical habitat drove these species to extinction.

Island Species are especially vulnerable. Today islands represent only 5.3% of our planet's dry land, yet about 75% of the extinctions recorded for amphibians, reptiles, birds, and mammals have happened on islands. The vast majority of these losses stem from habitat degradation by people or introduced animals—especially goats, pigs, dogs, cats, rats, and mongooses.

Species: Norfolk Island Kaka, *Nestor norfolcensis* (= *N. productus*)

Homeland: Norfolk Island in the South Pacific

Extinction Date: 1851

Cause: hunting by British convicts in a penal colony on this remote island (1788-1853); forest clearing

Image Credit: Bookplate by Dutch ornithologist/artist John Gerrard Keulemans (1842-1912)/ public domain; published in *Extinct Birds* (1907) by Lionel Walter Rothschild (1868-1937)

Species: Japanese River Otter, *Lutra lutra whiteleyi*

Homeland: Island of Japan; its population was once in the millions

Extinction Date: 1990s

Cause: over-hunting for otter fur, habitat destruction, and pollution

Commemorative Stamp: Issued by Japan in 1974

Species: Thylacine (= Tasmanian "Tiger"), *Thylacinus cynocephalus*—a shy carnivorous marsupial. Both sexes had a pouch —the male's pouch protected his genitals—reported in only one other marsupial, the water opossum.

Homeland: Island state of Tasmania; mainland Australia; New Guinea

Extinction Date: 1936, officially declared in 1986

Cause: competition from sheep ranching with bounty hunting between 1830-1909; reseachers have shown that disease was not a major cause.

Image Credit: painting by artist Henry C. Richter; published in John Gould's 1863 edition of *Mammals of Australia*/public domain

Species: Greater Akialoa (= Oahu Akialoa), *Akialoa ellisiana*

Homeland: Oahu, Hawaii, USA

Extinction Date: 1940

Cause: Historically, more than 30 species of Hawaiian honey-creepers have been wiped out by Polynesian and European settlers. They destroyed wildlife habitats and introduced mosquitoes to the islands in 1826. Mosquitoes carry avian malaria and only nine species of honey-creepers have survived—most now Threatened with extinction.

Image Credit: By Dutch ornithologist/artist John Gerrard Keulemans (1842-1912), published in *The Avifauna of Laysan and the Neighbouring Islands*, 1893

Most island species have evolved in the absence of mainland predators, so they often lack fear and defenses against strangers. Typically, island animals also take longer to reach sexual maturity and tend to invest energy in fewer offspring that are larger in size. Larger young are less vulnerable to small native predators and extreme weather and are stronger competitors for limited resources. In contrast, mainland environments favor (through natural selection) fast growth, earlier maturity, and a large number of smaller offspring to offset heavy predation. Most island species also have lower genetic variation than related mainland species, which reduces their ability to adapt to change.

Extinction rates of the world's flora and fauna continue to climb as human populations grow. Simply put, Planet Earth is much like a nutrient-filled Petri dish that has been inoculated with bacteria. A super-strain with the capacity to overtake all other strains in the dish will spread until it has gobbled up what remains of these life-sustaining nutrients. And when it reaches the edges of the dish, guess what happens? No living creature has ever been able to overexploit its environment without crashing. Perhaps it can jump to an uninhabited Petri dish with just the right milieu for survival, but that's no more likely than *Homo sapiens* finding another planet as friendly as the one we know and love.

Assuming that we do love our planet, it's time to recognize that we are headed to the edge of our Petri dish. In Earth's 4.5-billion-year history, no single species has had the capacity to control the fate of so many other life forms. We alone have the tools and ingenuity to predict and forestall what lies ahead, if we are smart enough to use them. But our species seems to prefer dealing with crises after they surface. This is one time when waiting for disaster to knock on our door might be our last perilous blunder!

Lonesome George: the last Pinta Island Tortoise, *Chelonoidis nigra abingdonii*—a "poster child" for species in trouble

Homeland: Pinta Island, Galápagos Archipelago, Ecuador

Extinction Date: 24 June 2012

Cause: Seafaring explorers, pirates, and whalers collected thousands of Galápagos tortoises for food—stashed alive in the hold of a ship, then slaughtered as needed. Later, introduced goats decimated the vegetation required by native species for food and cover. Lonesome George died of natural causes after spending 40 years at the Charles Darwin Research Station.

Image: by author/2008

Ecuadorian Banknote: extinct currency—after crashing in value during the 1900s; in 2000, Ecuador adopted the US Dollar as their official currency.

Species: Pyrenean Ibex (= Spanish Tur), *Capra p. pyrenaica*

Homeland: Pyrenees Mountains of Andorra, Spain, & France

Extinction Date: 6 January 2000

Cause: multiple factors— trophy hunting, disease, and competition from domestic livestock for food and habitat

Image Credit: painting by German natural history artist Joseph Wolf (1820-1899)/ public domain; from Richard Lydekker's book *Wild Oxen, Sheep & Goats of All Lands, Living and Extinct* (1898)

Species: Dodo, *Raphus cucullatus*, a fearless bird that could not fly

Homeland: Mauritius Island in the Indian Ocean

Extinction Date: 1688–1715, a recent estimate

Cause: Early Dutch sailors and settlers hunted them for food; its decline possibly aided by deforestation

Image Credit: © Naturmuseum Senckenberg/Sven Tränkner, Frankfurt, Germany

Oceans cover about 70% of our planet, contain 97% of Earth's water, regulate our climate, and support the greatest abundance of life we know. Marine algae and countless microscopic organisms called phytoplankton manufacture (through photosynthesis) more than half of the oxygen we breathe. When blue-green algae (cyanobacteria) started appearing in the fossil record about 3.5 billion years ago, they began infusing the ocean and Earth's atmosphere with oxygen. After that, the evolution of life on Earth was catapulted forward. We should be building monuments to honor these tiny photosynthesizers! And we should be paying closer attention to the health of our oceans.

A team of German scientists led by oceanographer Sunke Schmidtko has recently concluded that our oceans have lost more than 2 percent of their oxygen since 1960, based on 50 years of global data. Two percent might not sound like much . . . until one looks at what ancient sediments, rocks, and fossils have to say. Similar dips in dissolved oxygen over the past 200 million years correlate well with mass suffocation and extinctions of marine life. Geoscientists believe that volcanic activity triggered by continental drift spewed clouds of carbon dioxide into the atmosphere, which, over thousands of years, altered the temperature and chemistry of Earth's oceans. As water conditions became intolerable for most animal life, marine ecosystems collapsed. Burgeoning human population and run-away technology are catapulting our planet in the same direction at an unprecedented speed.

Our stewardship over the sea leaves much to be desired. So far, about half of our shallow-water coral reefs are gone, and scientists estimate that if conditions don't improve, most of the world's coral reefs could be lost in the next few decades. These reefs have been the life-line for 25% of all marine species on Earth—the primary supply of protein for billions of people worldwide. They are also an essential source of new medicines for this century,

a barrier that protects shorelines from erosion and property damage, and a priceless resource for recreation and tourism. The major causes of coral reef collapse are coastal development, sedimentation, pollution, invasive species, global climate change, and destructive fishing practices like dynamiting, use of poisons, overharvesting, and bottom trawling.

Living corals have a vital partnership with "algae" (*zooxanthellae*) that seek shelter within their tissues. These photosynthetic organisms make food that helps to sustain their host. They also give corals their color. But under environmental stress, such as water that is too warm, corals will expel their zooxanthellae—revealing the coral's white skeleton of calcium carbonate. This *coral bleaching* weakens corals, making them more susceptible to disease, which is spreading fast in places with warming seawater. Scientists are scrambling to find solutions and rescue corals before they disappear.

Acidification of seawater accelerates coral bleaching and affects a wide spectrum of other marine life. No less than 25% of the atmospheric CO_2 released by fossil fuel emissions is absorbed into the ocean. Once in the water, dissolved CO_2 forms carbonic acid. And water that is more acidic attacks protective shells made of calcium carbonate, including those of shellfish and microscopic plankton. Their shells become weaker and deformed. Oyster larvae, for example, are unable to make shells in acidified waters along the Pacific seacoast of the USA and Canada, producing massive die-offs.

If we continue on our present path, coral reefs and much of the marine life we depend on are likely to disappear by the middle of the 21st century. Paleontologists who have examined climate change in the fossil record have noted that coral reefs have recovered after downfalls brought on by warming seas and high levels of CO_2 in the atmosphere—but the process took *millions* of years!

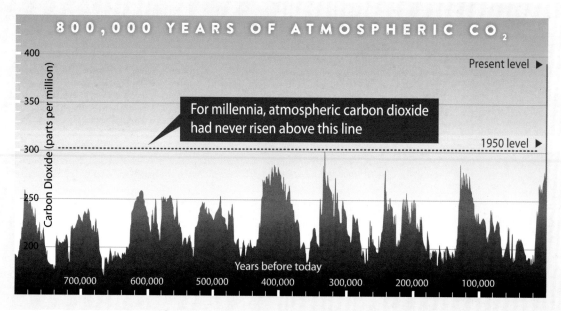

By analyzing ancient air bubbles trapped in ice, scientists have been able see how atmospheric gasses and climate have changed in deep time. Mile-thick ice core samples (and other paleoclimate evidence) show that fluctuating levels of the long-lived gas carbon dioxide (CO_2) have followed natural up-and-down cycles. But skyrocketing levels of atmospheric CO_2—set into motion around 1800 by the Industrial Revolution—are unprecedented and will continue to accelerate climate change in the unforeseeable future. Over the past 60 years, global atmospheric CO_2 has been increasing annually at a rate about 100 times faster than natural rates in pre-industrial times.

SOURCE: NASA GLOBAL CLIMATE CHANGE, A GLOBAL CLIMATE MONITORING PROGRAM

OCEANS IN TROUBLE

Buck Island Reef National Monument—a jewel of the US National Park System—is known for having some of the finest coral gardens in the Caribbean. At times, hurricanes can rejuvenate a coral reef, but frequent and severe storms will reduce a thriving reef to an uprooted pile of rubble. Broken pieces of healthy coral can often re-grow, but not if buried, sand-blasted by strong currents, or suffocated by opportunistic algae—warming seawater favors algal growth. In 2009, the author captured this scene at Buck Island about a year after Category-4 Hurricane Omar ravaged this reef in the US Virgin Islands.

SEA SICK

As oceans get warmer and more acidic, coral bleaching is spreading like a global pandemic—as seen on this Caribbean reef in Curacao, photographed in 2010 by Barry Brown. Bleached corals can sometimes recover, but stressed corals are more vulnerable to disease, repeated bleaching, and physical abuse from extreme weather or human activity. A recent long-term study of 315 coral reefs in Micronesia found that under disturbance-free conditions, a coral community needs at least 9-12 years to fully recover after a severe bleaching event. Keep in mind that coral bleaching is a relatively new historical phenomenon—the world's first mass bleaching happened in 1998. And since then, stress on coral reefs has grown in scope and severity.

Diseases deadly to corals are on the rise too. Biologists have given them names—such as black-band disease, yellow-blotch disease, white-plague, and rapid wasting—but these new diseases are even more mysterious than coral bleaching. Their effects are clear, but figuring out what they are and how to stop them has remained elusive.

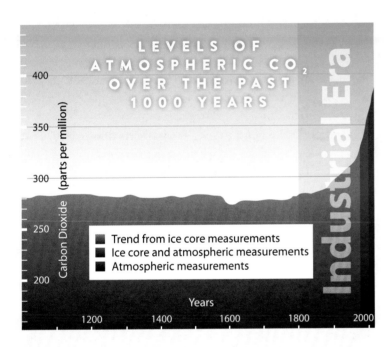

A closer look at the last 1000 years clearly shows the relentless rise of atmospheric CO_2 since the start of the Industrial Revolution. These are global averages. Note the close alignment of data derived from ice-core samples and direct atmospheric measurements. The dramatic climb in CO_2 during the past 200 years is directly correlated with fossil fuel burning.

Source: This graph is based on data presented in the 2007 and 2014 Synthesis Reports on Climate Change, prepared by the Intergovernmental Panel on Climate Change/Geneva, Switzerland. The IPCC "reviews and assesses the most recent scientific, technical, and socio-economic information produced worldwide relevant to the understanding of climate change." This organization does not conduct research—its work is policy-relevant and yet policy-neutral.

Bleaching elkhorn and brain corals in the Caribbean, photographed in Curacao by Barry Brown. Note the cap of algae/seaweed on this brain coral. Many common seaweeds release toxic chemicals that will bleach living corals if they make contact. Overgrowths of algae can also smother corals or kill them by blocking the sunlight they need to survive. So herbivorous fish that graze on seaweed play a vital role in keeping coral reef communities stable and healthy.

Our future is hanging by a thread. As individuals, the best thing that we can do to forestall a global catastrophe is to elect progressive decision-makers who will fight for sustainability, starting at the local level. National and international socio-economic priorities must change. Allegiance to self, party, church, or state must come second to values that offer hope for the future of our species.

The time has come for all of us to plan for and adapt to 21st-century environmental realities. Sea levels will continue to rise, shortages of food and fresh water will develop, and natural disasters will intensify. Decades ago, Scottish landscape architect Ian McHarg (1920–2001) recognized that urban development is rarely in tune with the way natural ecosystems work. In his now-classic book *Design with Nature* (1969), he focused on interdisciplinary planning for safer and more enjoyable living. Using maps with transparent overlays, McHarg laid the foundation for modern Geographic Information Systems (GIS). GIS computer programs can integrate layers of spatial data—such as roadways, floodplains, housing projects, and greenspaces—for sustainable community development. Visit the Ian L. McHarg Center for Urbanism and Ecology website.

Right now, in keeping with McHarg's principles, and anticipating rising sea levels, some city and regional planners are re-thinking coastal management practices. In the USA, a team of ecology-minded urban planners is re-designing Boston's coastlines. Check out their *Climate Ready Boston* website—it's an impressive model of what could and should be done for city preparedness. Their master plan integrates oceanography, commerce, transportation, sustainable building initiatives, renewable energy, health concerns, at-risk neighborhoods, and parklands for wildlife and public recreation. Using cost-benefit analyses, they have clearly shown that replacing concrete buffers and pavement with floodable greenspaces is their best long-term solution for flood-control.

Along the southwestern coast of the Netherlands, Dutch engineers have created *De Zandmotor*, the "Sand Engine." This man-made sand peninsula has been designed to protect nearby cities from the sea, while giving residents a fabulous recreational area. The project has been conceived to work *with* and not fight natural processes. If all goes as planned, the Sand Engine's huge mound of dredged sand will eventually disappear, rearranged by ocean currents, into a 12.4-mile-long (20-km) buffer that will shelter this coastline for the next two decades.

The more we let technology replace our respect for the interconnectedness of all living things, the more vulnerable we will be. In the years ahead, will our dependence on computers be a blessing or a curse? How many of us have even thought about being without power, fresh water, supplies, and communication networks for weeks, months … or years? Imagine the global consequences of just a threefold increase in the frequency and severity of natural disasters. How far can we go before reaching the tipping point? There *is* a threshold at which world economic and social order will quickly unravel.

Paleobiologists estimate that mass extinctions of prehistoric life have taken 5,000 years or more to play out—longer than the written history of humankind. But never before has one species set the stage for global ecological collapse, and we did it in less than 300 years, a blink in the eye of geological time. The more we disrupt our biosphere and the ecosystems that have made our planet a friendly place to live, the more likely it will be that humanity will experience a **sixth mass extinction** in a relatively short time frame.

Some ecologists call this a pending ***Anthropocene extinction***—doubly appropriate because *Homo sapiens* is not only bringing it on but will be among the losers. Rest assured, life on Earth will continue, with or without our species. Time-tested survivors—like cockroaches and crocodiles—won't save us. We are the only ones with the awareness and knowledge to protect our planet … and the future of today's youth. So it's our *shared responsibility* to act NOW!

GLOBAL WARMING
Photographed on a beach in North Andros, Bahamas.

MOTHER NATURE

Mother Nature has no favorites. She doesn't care if you are black, white, or blue. She doesn't care if have two legs or ten. She doesn't care if you worship a tree, a god, or your navel. She doesn't care if you live in a palace or under a rock. She doesn't care if you saunter, swim, or slither. Her indifference envelops every living creature on Planet Earth. If we—self-proclaimed brainiest of the bunch—foul our nest, this mother will not rush to our rescue. Our only assurance is that life in some form will go on, with or without us.

Over the past 48 years, I have spent a lot of time on islands, mostly in the Caribbean. For three years, I worked as an ecologist on the remote island of Mona—just over the western horizon from Puerto Rico. There I experienced a Category 5 hurricane and saw Mother Nature's indifference at work, almost daily. She tested me, equipped me with survival skills, and stoked my imagination. Ships that passed by Mona at night flushed oily wastewater into the sparkling sea—they left their mark on the island's coast and wildlife. I once found a doll's head that had washed ashore with a hardened hairdo of oil and sand. This "tar baby" was a perfect expression of Mother Nature's warning to humanity—foul your environment and you shall be among the victims. This was the beginning of a personal photographic art project that has grown over the years—I've included two samples on this page spread.

SEA DUMP

Photographed while scuba diving under a pier in Curacao, Netherlands Antilles.

SEEING THE UNSEEN

8

Whether peering into the depths of outer space or into a drop of pond water, mankind has always had the desire to look beyond what the unaided human eye can see.

Back in the Middle Ages, Arabs shaped and polished transparent beryl stones to make a simple hand-held magnifying lens for reading manuscripts up-close. And by combining two magnifying lenses into one device, Dutch spectacle maker Hans Janssen and his son Zacharias built the first known **compound microscope** in 1595. It could magnify objects 3-9 times bigger than can be seen with the naked eye. Since then, further refinements to light microscopes by scientists working in different countries have enabled us to see bacteria and the contents of small living cells, requiring magnifications of 400-1000x. But even the most powerful light microscopes cannot "see" viruses.

POLLEN GRAINS

Scanning electron microscope image of pollen grains (750x magnification) from an assortment of common plants, including oriental lily, geranium, sunflower, peony, dill, and *Epiphyllum* cactus. Image courtesy of Dartmouth College Electron Microscope Facility.

Some paleontologists specialize in studying tiny fossils, called microfossils, such as pollen grains and foraminifera. Most foraminifera (*forams* for short) are microscopic single-celled marine organisms with complex shells, typically about the size of the period at the end of this sentence. Forams are among the most abundant shelled organisms in the sea—there are over 4000 living species today and no less than 60,000 extinct species, dating back more than 500 million years! The structure of their shells is affected by water temperature and chemistry. Dying foraminifera continuously accumulate on the sea floor and their shells remain preserved as fossils in the sediments. By drilling into the sea floor to examine layers of forams, scientists have been able to better understand the evolution of Earth's oceans and climate.

Pollen grains—capsules that contain male sex cells of most land plants—are even smaller than forams. They average 50 microns (i.e. *micrometers*) in diameter. The width of an average human hair is about 75 microns, and the largest pollen grains are about 100 microns (= 0.1 mm) across. Being so tiny and not as charismatic to most paleontologists as macro-fossils like dinosaur bones and petrified wood, pollen is easily overlooked and has received relatively little attention in the fossil record.

But pollen has great stories to tell, best studied with a **scanning electron microscope** (SEM). SEMs use an electron gun to fire a high-voltage beam of electrons onto or through an object to create an image. As a scan progresses, electron-generated scanning lines, similar to those that produce the picture on your TV screen, are computer-compiled to create an image of the object. SEMs can precisely analyze an area as small as one square micron. The wavelength of the electron beam determines its resolution, and powerful SEMs can even reveal the structure of individual molecules and atoms.

The outer wall of a pollen grain contains one of the most durable substances found in the natural world. So it's not surprising that these tiny decay-resistant time capsules have persisted in the fossil record as far back as the Paleozoic when seed ferns colonized land. As in most modern plants, wind, water, or animals transport pollen to the plant's female organs. The pollen grain then germinates to form a tiny tube through which sperm can travel to fertilize an egg, starting the development of a seed. Plants can be classified by examining the shape, texture, and ornamentation of pollen grains. So studying pollen fossils can be valuable indicators of climate change and plant communities that are long gone.

While most pollen ends up within 100 m (3 football fields) from its source plant, some pollen can travel great distances. This is important to keep in mind when reconstructing plant communities of the past. Pine pollen, for example, has "wing-like air bags" designed for wind travel, up to 1,800 miles (2,900 km) in a short period of time. In a world with growing interest in and concern about genetically modified crops, we now know that pollen carried on the wind from a genetically engineered grass can pollinate other grasses up to 13 miles (21 km) away.

FORAMINIFERA

Below: Calcium carbonate shells of ten species of modern marine foraminiferans. Think of living forams as "amoebas with a shell." Many live on the sea floor (or lake bottoms); others float in the water column. To clearly show shell details, this black-on-white image was inverted to white-on-black. Photomicrograph (5x magnification) by Randolph Femmer/USGS (public domain). Unusually large forams (not shown here) reach 7 inches (18 cm) in size.

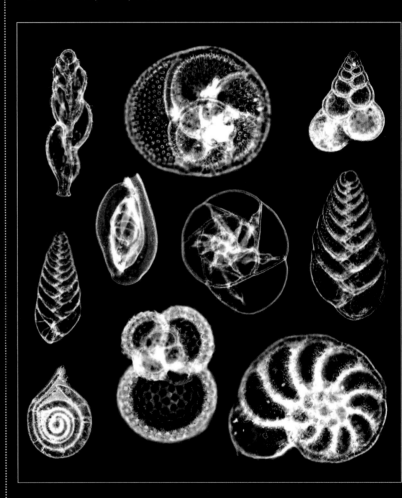

ERNST HAECKEL ART

Opposite page: German biologist-artist-physician-philosopher-professor Ernst Haeckel (1834-1919) was one of the most influential thinkers and talented artists in his day. A compound microscope aided his studies of tiny marine creatures, as featured in this 1904 book plate of Foraminifera (= Thalamophora). Clearly, a superb scientific illustration can reveal details unseen in a photo. Provided by the US Library of Congress Prints Division (LC-DIG-ds-07540).

Thalamophora. — Kammerlinge.

What we see is determined by the wavelengths of light that our eyes can detect. Global **electromagnetic radiation** covers a broad spectrum of wavelengths, measured in *nanometers* (nm). Only wavelengths in the 400–700 nm range are visible to the human eye. When these light wave frequencies are blended, as normally perceived, we see the world under "white light." And when white light passes through a prism, or water droplets that make a rainbow, these wavelengths are separated (i.e. *refracted*) into a spectrum of colors we see as red, orange, yellow, green, blue, and violet. We can also see unsaturated colors like pink, aqua, and brown.

The retina of our eyes contains four types of **photoreceptors**. Color-sensitive cells called cones are designed to detect reds, greens, and blues, and pass this information to the brain for processing. Other cells, the rods, are not sensitive to color—they can detect low levels of light and help with our night vision. But many animals see the world much differently than we do. Bees, jumping spiders, birds, and some rats, for example, can see in the *ultraviolet*, 10–400 nm wavelengths that are invisible to us. Some flowers have UV patterns on their petals that serve as "nectar guides," which to a bee would look like airport runway lights.

In fossil specimens, seeing and identifying features other than bone, such as muscle, internal organs, scales, or feathers can be challenging. And to make them visible for photography,

scientists have used side-lighting (cross-lighting), camera filters, polarizers, and more recently, **ultraviolet light (UV)**. Because different minerals in fossils glow (*fluoresce*) differently under UV light, unseen details of a specimen that are not evident in normal white light may become visible. UV light is especially useful for detecting restorations or repairs to a fossil—painted and repaired parts become easy to spot.

Three parts of the UV spectrum are used to examine fossils: UVA, UVB, and UVC. What we think of as "black light" is UVA radiation. Shorter UV waves are classified as UVB, and even shorter, UVC. Fortunately, UVC is naturally blocked by Earth's atmospheric ozone layer—it is the most harmful to living tissues. But when harnessed, UVC can provide a chemical-free way to disinfect water and sterilize surfaces. In paleontology, photographing fossils under UVA and UVB light is useful but is limited by its intensity and the variety of fluorescence that can be detected.

Specimens that remain dark under typical UV can be photographed fluorescing with clear detail using laser-stimulated fluorescence (LSF), a "next generation" analytical tool. When lasers excite molecules in minerals of a fossil, the elements will fluoresce, making them easy to distinguish from each other. LSF utilizes different wavelengths of light combined with optical filters to improve contrast. Which wavelength is most effective depends upon the specimen and the minerals in

Fossilized shrimp, *Aeger spinipes*, as seen under normal light (**above**), compared with a photograph of the same specimen under ultraviolet light (UVA). Ultraviolet light reveals glue used to restore the plate (blue) and paint applied to the shrimp (orange) to enhance its beauty/salability. Little of this fossil is real. In contrast, a similar specimen seen under UVA light (**right**) reveals that most of this shrimp is real—its exoskeleton glows white. Parts that don't glow, such as the antennae, have been painted. Specimens of Upper Jurassic age, from Solnhofen limestone at Eichstaett, Bavaria, Germany, courtesy of Martin Goerlich/ Eurofossils. UV lighting provided by Bruce & Rene Lauer/Lauer Foundation.

it, requiring some experimentation for best results. Translucent fish scales, for example, can be seen clearly enough to count growth rings. And LSF can reveal skin and other soft tissues in a fossil that are invisible under white light. Powerful lasers give the brightest florescence. LSF can even penetrate a short distance into the rock matrix of a fossil and reveal structures buried below the surface. The laser beam can be focused on microscopic fossils or used to scan larger specimens.

In the field, LSF has proven to be a useful tool for discovering fossils. Finding microfossils, for example, typically requires laborious screen-sifting, washing, and sorting through sediment. But by using LSF, these tiny fossils can often be spotted quickly and, remarkably, the process has even been automated. And because this tool is compact and portable, imaging specialist Thomas Kaye has been able to equip drones with LSF equipment to scan landscapes for fossil deposits. LSF technology also offers promise for detecting unusual mineral deposits and signs of life on other planets.

Below: A superb specimen of a predatory fish, *Aspidorhynchus*, with a flying reptile—a pterosaur, *Rhamphorhynchus sp.*—fossilized in Solnhofen limestone from Germany. Paleontologists and imaging specialists who have studied this specimen under UV light and with laser-stimulated fluorescence (LSF) have attempted to piece together this Jurassic drama. They propose that the fish had snatched the pterosaur from the air as it skimmed the water to catch small fish—like some modern bats and birds do today. And viewed under UV, a fish was found lodged in the reptile's throat. As predator and prey struggled, possibly they sank into a low-oxygen "dead zone" and died together— speculative, yes, but possible. LSF image (below) courtesy of Tom Kaye/Foundation of Scientific Advancement. Some details seen under LSF are not evident under white light— the red coloration, for example, suggests the presence of soft tissue. The author photographed the specimen under white light in Granada Gallery/Tucson, AZ, part of a private collection curated by Interprospekt.

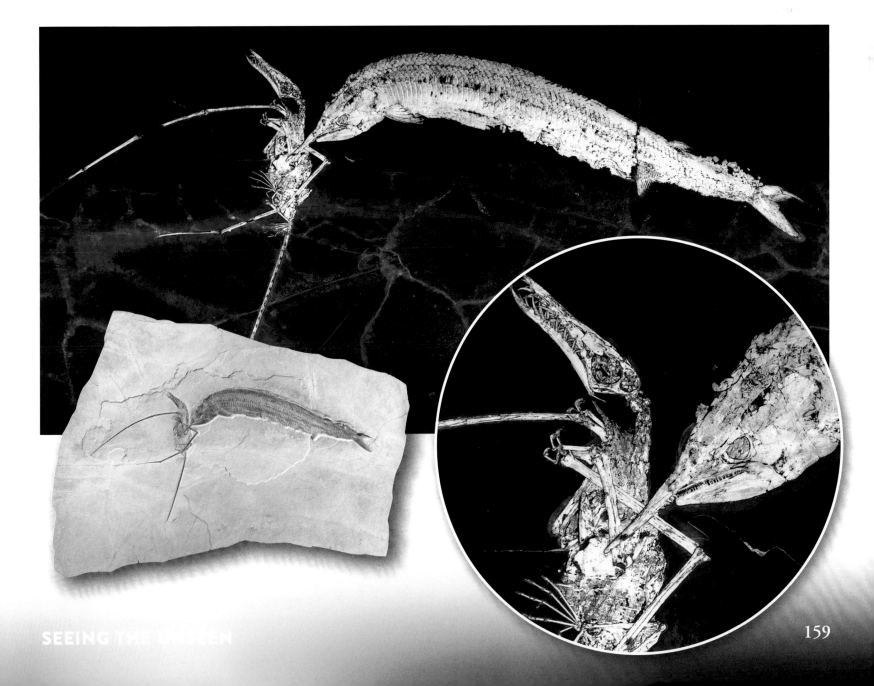

More than a century ago, medical doctors began using a new device for examining the human body with X-rays. This high-energy form of electromagnetic radiation can pass through most objects, and the resulting slice of information on exposed film allowed doctors to look inside the body without surgery. But it wasn't until 1969, with the aid of computers, that soft tissue details could be captured. This was the birth of **X-ray Computed Tomography (= Computerized Tomography or CT scanning)**, formerly known as **Computerized Axial Tomography (CAT)**. CT scans combine a series of X-ray slices taken from different angles to generate 3-D images. Modern computer hardware and software have improved scan speed and image quality, allowing us to reconstruct incredible 3D views inside living and fossilized organisms.

Needless to say, CT scanning has revolutionized paleontology. Specimens old and new are now being scrutinized from the inside out. To cite one example, in 1975 South African paleontologist James Kitching excavated a fossilized cast of a burrow system in 250-million-year-old Karoo sandstone. These ancient burrows are preserved not as tunnels, but as a plug of sediment that once filled them. Terrestrial creatures had presumably used these retreats for sleeping, to escape predators, and to avoid extreme weather. A portion of this particular trace fossil appeared to contain fossilized bone, so for safe-keeping, it was stored in the Evolutionary Studies Institute at the University of Witwatersrand in Johannesburg.

Thirty-seven years later, after the development of a state-of-the-art **European Synchrotron Radiation Facility** in Grenoble, France, a team of paleontologists decided to take a closer look at this lump of rock. Using a non-invasive CT scan, a remarkable discovery was made: the rock contained two complete, articulated skeletons of an entombed Early Triassic odd couple, a mammal-like reptile (a cynodont) and an injured temnospondyl amphibian. Detailed analysis of the orientation, condition, and posture of the two animals suggests that they died together in a flooded burrow. If we consider the behavior of living amphibians, it's not hard to imagine that these two species shared living quarters. Tiger salamanders, for example, often occupy rodent burrows, with no negative interactions between these animals.

MOLECULAR PALEONTOLOGY

Another revolution in paleontology is taking place on the molecular front. New techniques for extracting and analyzing molecules in tissue samples from fossils are offering a glimpse into the past that was never before possible. With increasing accuracy, biologists are better able to determine who is related to whom in the fossil record, what prehistoric creatures might have looked like, and how they might have behaved. This is a new and wild frontier in the evolution of our understanding of life on Earth.

Typically, complex molecules like DNA and proteins don't last long after an organism's death. Scientists predict that complete strands of DNA cannot survive more than 1.5 million years, even under ideal conditions. Frozen environments seem to be the most favorable. Genetic material from insects and plants found in Greenland ice cores is up to 800,000 years old. And DNA extracted from a leg bone of a horse frozen for 735,000 years in Canadian permafrost allowed scientists to reconstruct its full genome.

Finding "clean" DNA in fossils is one of the molecular paleontologist's biggest challenges. Typically, raw DNA extracted from bone or muscle in ancient specimens sits in a confusing soup of bacterial DNA. Researchers have found that hair offers a much purer source of ancient DNA. The hair shaft encases DNA in a protective keratin sheath, where it can better resist damage from the elements and contamination from bacteria. Sex hormones have also been extracted from Ice Age hair, which is beginning to provide insights into the physiology of woolly mammoths.

In 2016, Swiss biologists presented convincing evidence that both hairs and feathers evolved from scales of their reptilian ancestors. They determined that during early stages of embryonic development, the molecular and anatomical "signatures" of all three are identical.

As recently as 2007, many paleontologists believed that the colors of dinosaurs would always remain speculative, but in 2010, a team of British and Chinese scientists crushed part of that notion. They reported the first solid evidence for color patterns in feathered dinosaurs. Just as the sheath of a hair shaft protects its DNA from decay, the keratin sheath of a feather offers long-term protection for pigment cells within it. As in living birds, feathers in 125-million-year-old dinosaur fossils like *Sinosauropteryx* contain microscopic capsules of melanin, natural pigments common to most living organisms. In bird feathers, black and brown pigments are packaged in sausage-shaped capsules, while reddish pigment is seen in ball-shaped capsules. Areas with no pigment were assumed to be white. So by examining the well-preserved pigment capsules within fossilized feathers, this team of biologists was able to reconstruct color patterns in the feathered dinosaurs being studied. Nevertheless, dinosaurs without feathers will, at least for now, remain fertile territory for the imagination of artists.

CLEAN DNA

These woolly mammoth vertebrae with bits of hide and hair—found frozen in Siberia, Russia—offer a promising source of ancient DNA. A keratin sheath around the shaft of each hair protects its DNA from weathering and bacterial contamination. Thus hair offers a much cleaner source of DNA than bone or muscle. Photographed in GeoDecor's showroom with permission from owner Tom Lindgren.

HIDDEN WONDERS

Above: Paleoecologist Vincent Fernandez scans a lump of fossil-containing rock at the European Synchrotron Radiation Facility in Grenoble, France. This co-operative research facility is supported by 22 countries and is used by thousands of scientists. The rock being 3D-scanned with X-ray Computed Tomography (= CT scanning) is fossilized fill from an ancient burrow system in South Africa. Much to everyone's surprise, the scan revealed two animals sharing the same burrow—a meerkat-sized ancestor of mammals, *Thrinaxodon*, and an early amphibian, *Broomistega* (**inset**). To see an amazing 20-second video of these two skeletons rotating in 3D, as imaged by the synchrotron, visit https://fossilsinsideout.com/fossil-CT-scanning. Source: © 2013 Fernandez, et al.; https://doi.org/10.1371/journal.pone.0064978. Photo of Vincent Fernandez © Pascal Goetgheluck.

ACID ETCHING

Neither X-rays nor CT scans have been successful in revealing embryonic structures within dinosaur eggs—they deliver murky images. So in the 1990s, British scientist-technician Terry Manning decided to resurrect a fossil preparation technique that dates back to the 1930s. He used dilute acid solutions to dissolve the calcite and silt matrix in dinosaur eggs without harming the embryo's delicate fossilized bones. His painstaking process required multiple acid baths, cleaning with a needle and brushes, and protecting exposed bones with a clear preservative—steps that required several months to complete. The enlarged view (**left**) clearly shows the embryo's skull pointing downward in the jumble of tiny bones. Specimen and photo by Terry Manning.

THE ART OF
FOSSIL
PREPARATION & DISPLAY

9

In the business of paleontology, many passionate people who create order from chaos rarely get the recognition they deserve.

Those who dig, prepare, and exhibit fossils must have a mastery of both art and science. Much of what happens behind the scenes determines a specimen's value to the scientific community, collectors, and our understanding of the history of life on Earth.

FOSSIL PREP LAB

A typical day in Fossilogic's paleo prep lab. Everett Black (far left) wears a half-mask respirator while blowing limestone dust away from a turtle specimen. Owner Brock Sisson is welding a support armature for a skeleton. The two dinosaur skulls in the foreground are replicas—the smaller, tan one is a life-size cast of *Majungasaurus atopus*, a fossil from Majunga Basin in Madagascar; and the larger white skull, a work in progress, belongs to a new species of therapod dinosaur. Creating this sculpture from the few original parts found requires artistic skill combined with scientific understanding of similar skulls.

Because plant and animal remains are at the mercy of geological processes, fossils are usually found in pieces, often scrambled, cracked, crushed, discolored, and distorted. Some can be found by exploratory digging and cutting to remove fossil-bearing slabs from a rock quarry. But often a bone fragment found in an eroded hillside is where the rescue effort begins. Most surface fossils, in fact, are discovered in places where water and wind are exposing and destroying these prehistoric remains. Paleontologists strive to recover them before they are lost forever.

Excavating fossils from the ground can be as simple as digging them out with a hand shovel or as complex as using heavy equipment, power tools, and even explosives to expose them. Old extraction methods were often crude, resulting in many damaged specimens. In professional digs today, the location and relative position of each fossil or group of fossils is usually mapped and their removal is precisely executed. Individual bones are often numbered for reference. Preserving such information can be useful for future research, but the process is slow and labor intensive. Extracting a nearly complete dinosaur skeleton, for example, commonly requires a crew of five to ten trained people investing two or three years in seasonal digging.

Typically, fossilized bones are fragile and not easy to separate from the rocky matrix in which they are buried. So chunks of rock with bones in them are wrapped in a protective, stabilizing field jacket made from plaster strengthened with burlap or paper—similar to the casts that doctors use to mend broken arm or leg bones. Fossils can then be safely transported to a laboratory for careful cleaning, restoration, and evaluation. In museum and university collections, many fossils remain stored in their plaster jackets until specialists can find the time and funding to study them.

Henry Galiano and Brock Sisson assemble a skeleton of the dinosaur *Allosaurus fragilis* (**center**) of Jurassic age from the Morrison Formation in northern Wyoming. This is a mostly complete, articulated specimen— one of the best ever found—with much of its skeleton still supported by a plaster field jacket. Reconstruction in Brock's paleo lab, Fossilogic, Pleasant Grove, Utah. Photo © 2018 Fossilogic.

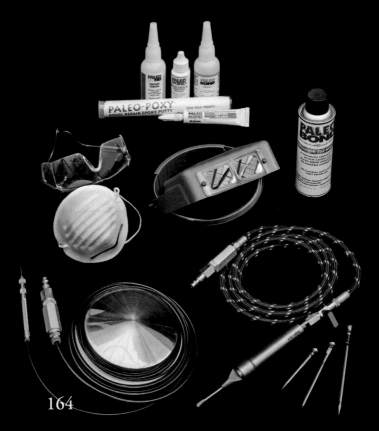

Tools and supplies commonly used in the preparation of paleontological specimens.

From top to bottom: glues and fillers for the restoration and stabilization of cracked or broken parts; safety glasses for eye protection and face mask to stop dust inhalation; optical magnification visor; and two pneumatic air scribes with styluses for precision matrix removal. The tool on the left is a "Pin Vise Puffer," often used when working on very small fossils under a microscope. Its shiny metal disk is a dental rheostat air foot pedal that allows the preparator to finely regulate airflow from the stylus to blow away debris. The tool on the right is the PaleoTools Super Jack, which operates much like a mini-jackhammer with precise control; it is equipped with interchangeable stylus points for different tasks. Tools and supplies borrowed from PaleoTools.com for this photo.

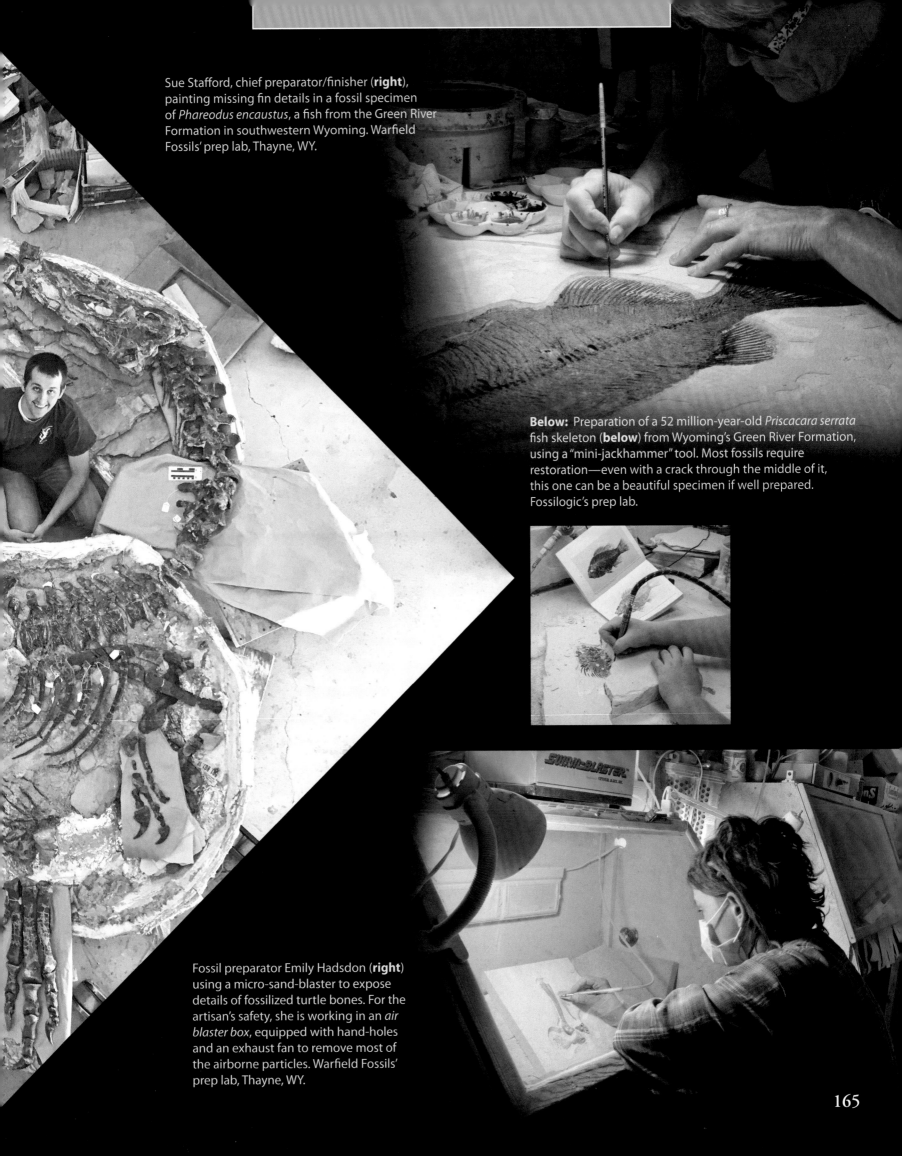

Sue Stafford, chief preparator/finisher (**right**), painting missing fin details in a fossil specimen of *Phareodus encaustus*, a fish from the Green River Formation in southwestern Wyoming. Warfield Fossils' prep lab, Thayne, WY.

Below: Preparation of a 52 million-year-old *Priscacara serrata* fish skeleton (**below**) from Wyoming's Green River Formation, using a "mini-jackhammer" tool. Most fossils require restoration—even with a crack through the middle of it, this one can be a beautiful specimen if well prepared. Fossilogic's prep lab.

Fossil preparator Emily Hadsdon (**right**) using a micro-sand-blaster to expose details of fossilized turtle bones. For the artisan's safety, she is working in an *air blaster box*, equipped with hand-holes and an exhaust fan to remove most of the airborne particles. Warfield Fossils' prep lab, Thayne, WY.

Fish, leaves, trilobites, insects, and other creatures commonly found in limestone quarries are seldom removed from the rock in which they were preserved. Layered limestone often splits to expose part of a fossilized plant or animal, but the break is seldom clean. Unwanted matrix must be carefully cleared away and any damage repaired. Wyoming paleontologist and quarry owner Rick Hebdon, who has excavated and prepared thousands of fish from the Green River Formation, estimates that only about 1 in 100 fish makes a clean split from the rock, requiring no further preparation. The rest must be cleaned with an arsenal of micro power tools, sometimes under a microscope. Broken pieces get glued, and many must be touched-up with artist paints. Rick claims, from decades of experience, that if you could count the fish being dug in all Green River quarries in a single season, the number would come close to 200,000. Commercially, there is a huge demand for small fish sold wholesale for a few dollars each—but the profit margin is slim.

Preparing fossils for study or exhibition requires knowledge of geology, anatomy, ecology, and structural design, along with technical skills, which can include the use of specialized hand tools and resins, microscopes, computers, scanners, and 3D printers. For many specimens, lab work begins with cutting off the plaster field jacket to expose the matrix containing the fossil(s). Fossils left in their matrix to minimize the risk of damage can be studied using new imaging technologies like X-ray tomography. But most fossils destined for collections or public display must be meticulously exposed, cleaned, and cared for.

Techniques for cleaning and preserving fossils are as varied as the fossils themselves. For example, bones retrieved from tar pits must be soaked overnight in a solvent to remove hardened asphalt. But in most cases, exposing bones or delicate parts requires the careful removal of rock and sediment with the aid of small chisels, brushes, and tools that appear to belong in a dental office. The harder the matrix, the more challenging the job. Face masks and glove boxes are often needed to protect these artisans from dust inhalation—limestone dust can cause silicosis, scarring of the lungs that can become cancerous. Visible cracks and imperfections in specimens are filled with various types of stabilizing glues and resins. To reconstruct a fully articulated skeleton, cleaned bones are arranged in anatomical order to determine how many are missing. Among dinosaurs, it's unusual to find specimens that are more than 50% complete. And very few have been found with 90% or more of the bones represented—these are of greatest scientific value.

A beautifully mounted skeleton of a "false saber-toothed cat," *Dinictis sp.*, of Oligocene age from White River Badlands, South Dakota. This lynx-sized cat belongs to an extinct family, Nimravidae, endemic to North America from the Late Eocene to Early Miocene epochs. It had incompletely retractable claws, powerful jaws, and a long tail.

Cleaning and restoring original bones and then casting missing parts to complete this skeleton required about 4-6 weeks of labor. The artistic challenge of mounting the skeleton in an engaging life-like pose took two people an additional 4-5 months. Working with small bones made this job especially challenging. Its legs and vertebral column are supported with high-carbon steel, and tiny set-screws hold the vertebrae in place. As is often the case with articulated skeletons, this one had to be designed for safe and convenient transport—every bone is easy to remove. And to finish the piece, an attractive hardwood base had to be crafted. Photograph and preparation of specimen courtesy of Brock Sisson/ Fossilogic.

Tyson Hunter preparing a huge limestone slab with a fossilized palm frond, *Sabalites sp.*, from Wyoming's Green River Formation near Kemmerer, WY. Fully exposing and restoring a fossil of this size requires months of work. Green River Stone Company's prep lab in Logan, Utah.

167

Specimens with missing parts can be reconstructed in several ways. Collecting sites often contain the skeletons of multiple individuals, so borrowing parts from others of the same species, size, age, and sex to make a "Frankenstein composite" is a common and fully acceptable scientific practice. Many museums have skeletal elements in their collections that can be borrowed and cast to fill gaps. Institutions and individuals with the best specimens often share what they have, and 3-D scanning/printing makes this both practical and economical today. But when all else fails, a sculptor can be hired to reconstruct missing bones. Mistakes are occasionally made when assembling composite skeletons, but as new finds surface, corrections are put into place. Collectors are responsible for recording and disclosing all aspects of the specimen's history to future owners.

Reconstructing *Patagotitan mayorum*, a gigantic sauropod dinosaur from Argentina that would dwarf *T. rex*, was accomplished using parts from at least six partial skeletons unearthed in one bone bed. About 30% of its skeleton had been lost in time, but the completeness of the other 70% made the assembly of this amazing skeleton possible. Missing pieces were reconstructed in Canada from 3-D scans of the Argentinian bones. Thanks to bilateral symmetry in vertebrate animals, bones can be replicated to make matching pairs. Using scans and state-of-the-art robots to carve molds from polystyrene, missing bones were cast in fiberglass. Sadly, the titan's skull was never found, so a substitute was modeled after those from other sauropods. If a skull of this species is found in the years ahead, casts on display in museums will be updated. Computer imaging of the fully articulated skeleton also aided the production of iron framework needed to hold the bones together. Overall, a team of 40 working for two years—an estimated 40,000 man-hours of time—were invested in the excavation and reconstruction of this skeleton! Researchers estimate that this colossus, a young adult, exceeded 120 feet (37 m) in length, matched the weight of 14 African elephants, and reached the height of a seven-story building.

Fossil replication has revolutionized paleontology. In fact, life-sized casts of this titanosaur are now on public display in Chicago's Field Museum and in New York's American Museum of Natural History. Many other museums, educational institutions, and traveling exhibitions showcase accurate replicas of fossils that are housed in far-away places or collections that are off-limits to the public. Most are as convincing and exciting to visitors as seeing the real thing. Casts are lightweight, easy to move, and economical to buy and own. They also give many paleontologists the option to borrow exact replicas of rare and fragile specimens for study.

Paleontologist Peter Larson measuring a *Gorgosaurus* dinosaur skull in a lab devoted to preparing realistic cast replicas of fossils at the Black Hills Institute of Geological Research in South Dakota.

Below: Storage facility for dinosaur bones in plaster jackets, awaiting preparation and study at the Black Hills Institute of Geological Research.

Staff member Rick Reed at the Rocky Mountain Dinosaur Resource Center uses a hand-held state-of-the-art 3D scanner to capture an image of the skull of a prehistoric aquatic reptile, a *Platycarpus* mosasaur. The scanner converts the skull into a virtual 3D object, data that can be output later ("printed") to produce a life-like replica. Even full skeletons can be scanned on-site without removing them from a museum.

Original skull of *Tyrannosaurus rex* named STAN (**top**)—the nickname given to this fossil—discovered by amateur paleontologist Stan Sacrison in South Dakota's Hell Creek Formation in 1987. The identity of these bones was not recognized until 1992, when Peter Larson visited the site. His team from the Black Hills Institute of Geological Research/BHIGR located all except two of the 50 bones that make up a T-rex skull—the best of its kind ever found. Preparing, casting, and mounting this specimen required several thousand hours of work. Note the elaborate framework needed to hold the pieces in place.

Besides its skull, most of STAN's skeleton was found too. The team at BHIGR cleaned the bones, molded them, and assembled this cast replica (**above**)—on display at the Rocky Mountain Dinosaur Resource Center, Woodland Park, Colorado. Cleaning and restoring the original bones and making the molds took at least 25,000 hours of work. This animal was 40 feet (12 m) in length, stood 12 feet (3.7 m) tall, and had a 5-foot (1.5-m) skull with 10-inch (25-cm) teeth.

Specimen buyers, beware of unscrupulous sellers—especially on internet sites like eBay, in gift shops, and at trade shows! Be suspicious and do your homework. Learn about your subject matter. Ask sellers about the origin of each specimen and whether they were involved in the preparation. And never hesitate to ask for advice from experts. Remember that most fossils require restoration to make them aesthetically pleasing, which is not the same as fakery. Honest dealers will tell you how the specimen was restored or modified from its original state. Even a cast is not considered a "fake" if it's advertised as a replica. Keep in mind, too, that a "Certificate of Authenticity" from a seller means nothing unless it is signed by a respected dealer who knows the specimen's **provenance** (its record of origin, ownership, and history). Consequently, most paleontologists and professional collectors deal only with sources they know and trust.

Trilobites are popular in the fossil trade, and the fabrication of fakes has been on the rise since the 1980s. Especially in Morocco, a few unscrupulous European and American dealers encouraged fakery for profit. Since then, fakery has become a flourishing industry that has risen to new levels of sophistication in Morocco, Russia, Bolivia, and China. There are more than 20,000 described species of trilobites. Parts from unrelated species are sometimes cobbled together to create convincing exotics, and others are skillfully crafted from colored resins mixed with pulverized stone. Real trilobites are seldom perfect and often have cracks or natural mineral seams that penetrate both the fossil and its base. Those cast in resin can be identified by examining them under UV light—glue and plastic parts will glow—or by inspecting them closely with a magnifying lens. Look for tiny holes or surface craters from broken air bubbles. And examine the eyes—most trilobites have intricately faceted compound eyes that are difficult to replicate precisely.

Amber is fossilized tree resin that has aged for at least 10 million years. And because this semi-precious stone is highly prized by jewelers and fossil collectors, realistic look-alikes have flooded the market. Artificial amber is being made from glass, an array of plastics, and young, unfossilized tree resins called **copal**. To make pricey fakes that can convince buyers of their authenticity, some artisans add insects and other inclusions to the mix. Once again, it is wise to do some research and detective work before buying amber from dubious suppliers.

Amber is lightweight, warm to the touch, and has a non-glossy sheen. Here are six useful tests for the imposters—note that more than one test might be needed.

Friction: Rubbing amber or copal in your hands will produce heat and smell like tree resin, not plastic. Vigorous rubbing on cloth for 20–60 seconds will generate an electrostatic charge in amber that attracts hair or dust particles. Copal gets sticky and won't carry a charge.

Taste: Wash a piece of amber in soapy water and taste it—most fakes have a plastic or chemical flavor, and amber has none.

Buoyancy: Genuine amber (and copal) will float in a mixture of ¼ cup salt dissolved in 2 cups of warm water—most fakes will sink.

UV Light: Amber, some plastics, and certain types of glass will glow under UV light, so anything that doesn't glow can be ruled a fake.

Acetone: A reliable amber vs. copal test can be done with a drop of acetone or nail polish remover. After it evaporates, amber remains unaffected. On copal, the spot might look whitish and will be sticky when rubbed with a tissue.

Heat: Plastics burned or poked with a hot pin will melt and smell like plastic. A hot pin will penetrate copal easily and amber with difficulty. Test this on an unimportant part of a specimen.

Note that insects in real amber are blackish or brownish, not colorful, and never outlined. Furthermore, few insects in amber look perfect. Copal that has aged for several hundred or a few thousand years may take a beautiful polish, but beware—it's not fossilized and will craze on the surface as volatile chemicals in the resin evaporate in the years ahead.

Like many fish fossils, some fossil plants and invertebrate animals are extremely abundant where they are found, allowing secondary art markets to grow. These raw materials are being fashioned into stunning jewelry, lamps, tables, tiles, framed mosaics, and elaborate sculptures. When cut and polished, coiled ammonites and agatized wood are especially popular among artisans. Their patterns and colors are eye-catching and can transmit light beautifully.

Above: pieces of raw amber and copal—often difficult to distinguish with the naked eye.

Left: bee, mosquito, and spider in Dominican amber—courtesy of the Amber Museum, Dominican Republic.

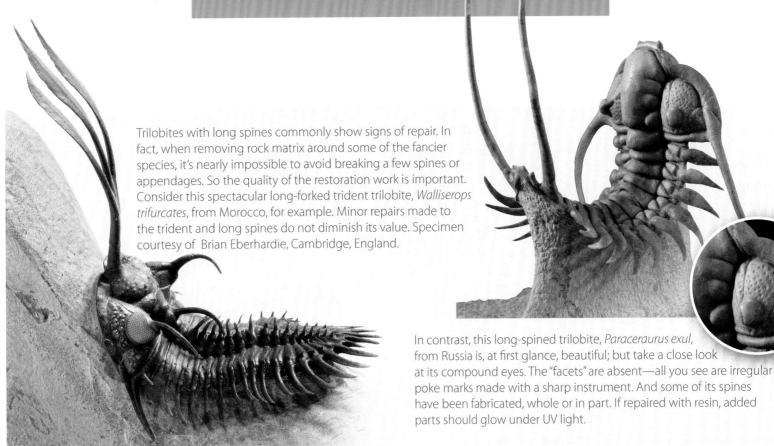

Trilobites with long spines commonly show signs of repair. In fact, when removing rock matrix around some of the fancier species, it's nearly impossible to avoid breaking a few spines or appendages. So the quality of the restoration work is important. Consider this spectacular long-forked trident trilobite, *Walliserops trifurcates*, from Morocco, for example. Minor repairs made to the trident and long spines do not diminish its value. Specimen courtesy of Brian Eberhardie, Cambridge, England.

In contrast, this long-spined trilobite, *Paraceraurus exul*, from Russia is, at first glance, beautiful; but take a close look at its compound eyes. The "facets" are absent—all you see are irregular poke marks made with a sharp instrument. And some of its spines have been fabricated, whole or in part. If repaired with resin, added parts should glow under UV light.

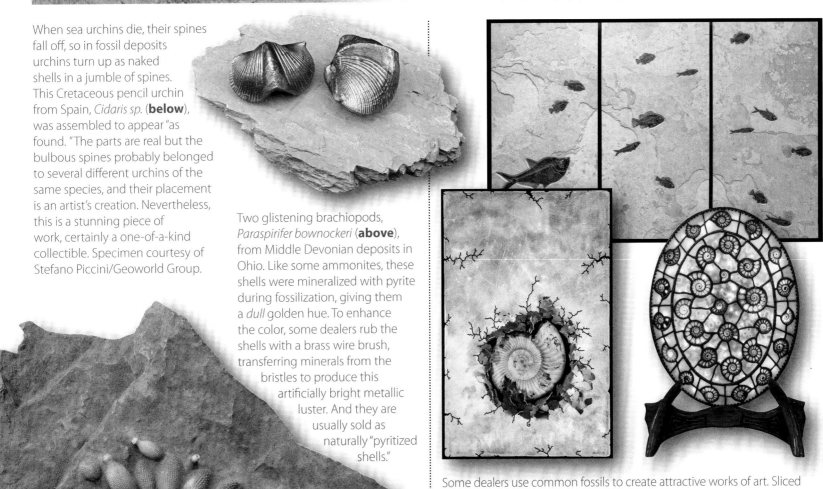

When sea urchins die, their spines fall off, so in fossil deposits urchins turn up as naked shells in a jumble of spines. This Cretaceous pencil urchin from Spain, *Cidaris sp.* (**below**), was assembled to appear "as found." The parts are real but the bulbous spines probably belonged to several different urchins of the same species, and their placement is an artist's creation. Nevertheless, this is a stunning piece of work, certainly a one-of-a-kind collectible. Specimen courtesy of Stefano Piccini/Geoworld Group.

Two glistening brachiopods, *Paraspirifer bownockeri* (**above**), from Middle Devonian deposits in Ohio. Like some ammonites, these shells were mineralized with pyrite during fossilization, giving them a *dull* golden hue. To enhance the color, some dealers rub the shells with a brass wire brush, transferring minerals from the bristles to produce this artificially bright metallic luster. And they are usually sold as naturally "pyritized shells."

Some dealers use common fossils to create attractive works of art. Sliced ammonites, for example, transmit light like stained glass—the lamp shown here (**above right**) was designed by Konstantin Soyfer, a Tucson artist from the Republic of Moldova (formerly Soviet Moldavia). Artist Hendrik Hackl, based in Mannheim, Germany, created this modern wall sculpture from brushed aluminum and an ammonite fossil that he personally cleaned and prepared for display (**above left**). The Green River Stone Company puts the abundant fish found in their Wyoming quarry to good use by artistically embedding them in panels of rock suitable for home or office décor (**top**).

171

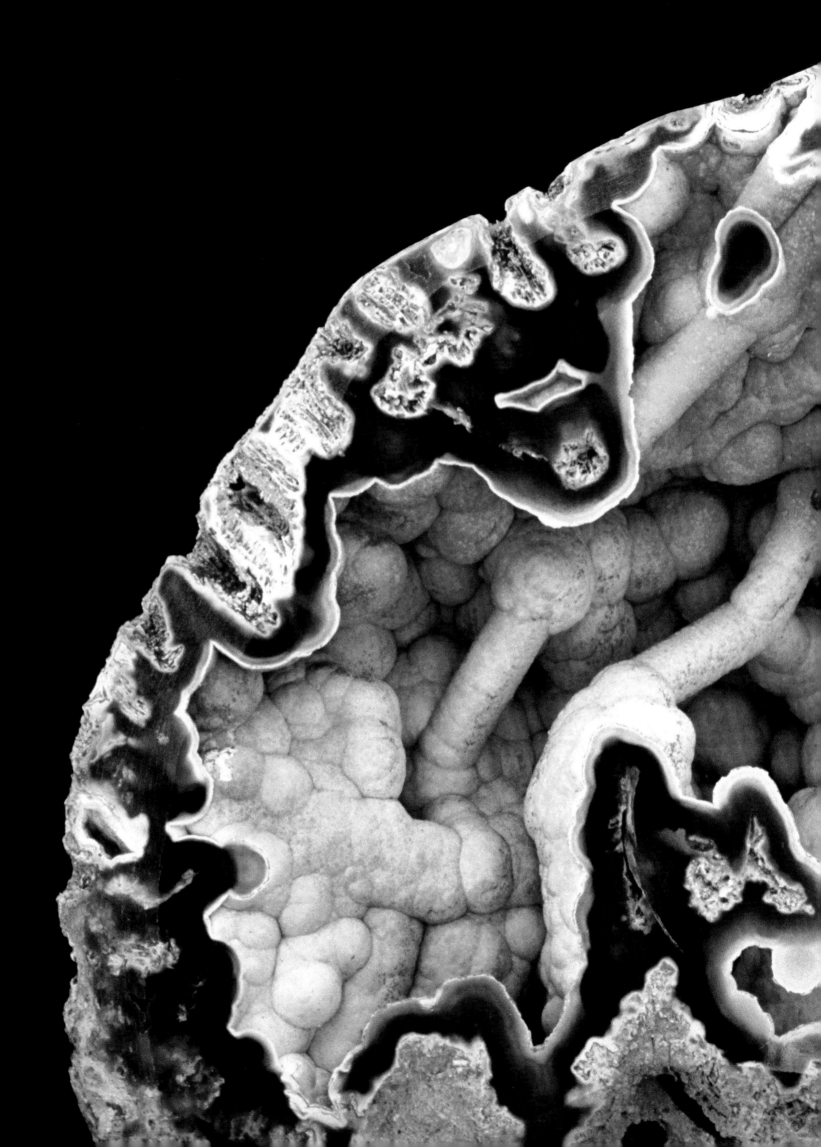

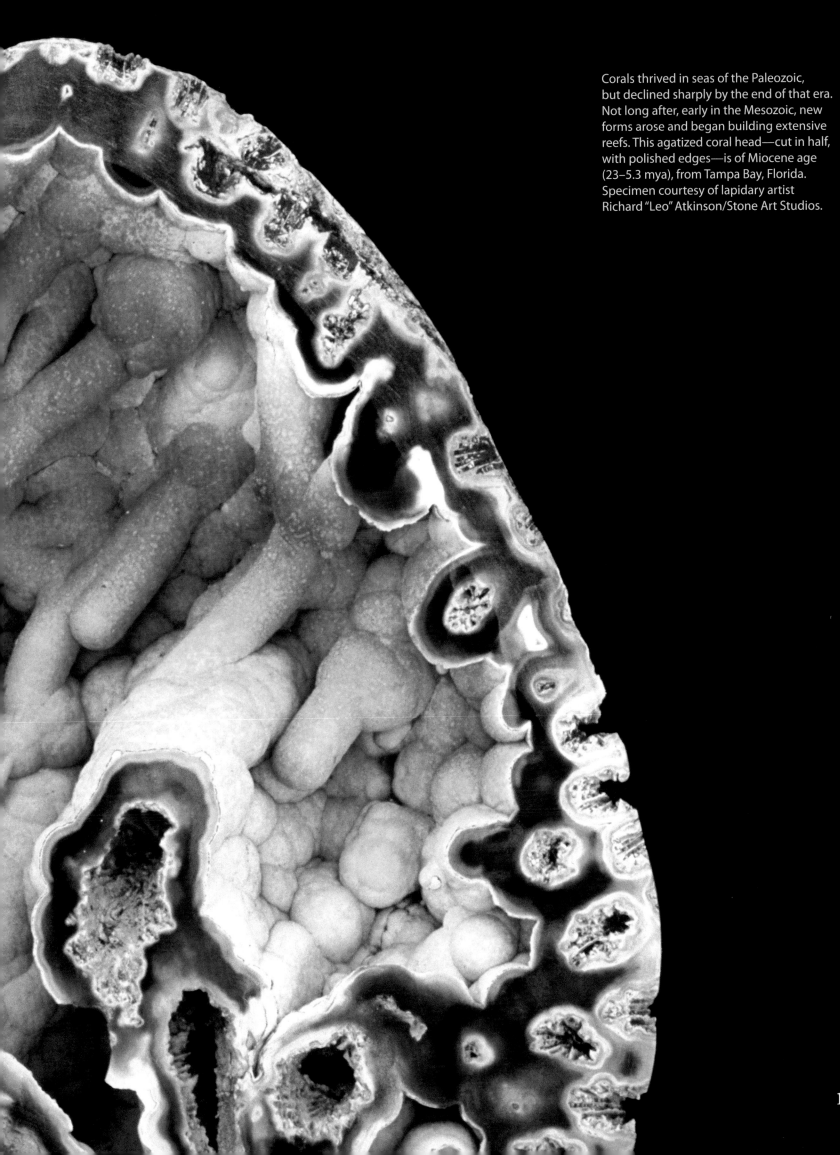

Corals thrived in seas of the Paleozoic, but declined sharply by the end of that era. Not long after, early in the Mesozoic, new forms arose and began building extensive reefs. This agatized coral head—cut in half, with polished edges—is of Miocene age (23–5.3 mya), from Tampa Bay, Florida. Specimen courtesy of lapidary artist Richard "Leo" Atkinson/Stone Art Studios.

Visitors to Utah's Museum of Ancient Life will experience many of the superb exhibits designed, fabricated, and installed by Chase Studio (see text). Some of the exhibits are meant to be touched—walk in a forest of dinosaur skeletons for example (**above**). Or peer into a glass case with one-way angled mirrors that provide multiple reflections for depth and spaciousness (**left**). The case contains an underwater "garden" of Early Cretaceous sea life, no less than 30 species—flower-like sea lilies (crinoids), sea stars, sea urchins, brachiopods, snails, trilobites, cone-shaped bryozoans, solitary corals, and more. It's hard to believe that this modern, state-of-the-art exhibit was installed in 2001!

MUSEUMS

Exploring a great natural history museum can be a life-changing experience. As an ecologist and former tour organizer, I have visited many museums, zoos, aquariums, and botanical gardens worldwide. Two years ago, thanks to paleontologist and fossil restoration artist Brock Sisson, I discovered a little known museum tucked away in the town of Lehi, Utah, about an hour's drive south of Salt Lake City—the Museum of Ancient Life at Thanksgiving Point. This relatively small museum is filled with world-class exhibits, many of them created and installed about two decades ago by Chase Studio, located in Missouri's Ozark Mountains.

Lurking in the shadows behind every superb exhibit is a close collaboration between scientists and highly skilled artists and artisans. These are the people who bring the past to life. While

writing this book, I spoke with owner Terry Chase to find out more about the business of exhibit design. Terry, who founded Chase Studio in 1973, is an artist-scientist with a PhD in invertebrate paleontology. He and his talented staff of 90 have constructed several hundred paleo dioramas for leading museums of the world. Their exhibits represent life forms from every geological period, supported by a private reference collection of over one million specimens, most of which are fossils. The Smithsonian donated 500 metal-drawer cabinets to store this collection safely. Chase Studio has a large graphics department, mural painters, a taxidermist, model builders, and a shop with cabinet makers, welders, electricians, and lighting experts. Their awe-inspiring exhibits are all built in-house, in their Missouri studio. Fossil replicas are cast from a wide variety of polymers, ranging from urethane to polyethylene. Models are hand-painted, usually with an airbrush, and in some cases the casting resin is tinted for a translucent look. As of 2020, they had not used 3D scanners and printers.

Museums ignite curiosity, expand awareness, spark imagination, and foster critical thinking. Today especially, children need engaging opportunities to explore beyond their electronic devices. Expansion of technology with rising use of smart phones has been a blessing and a curse. Researchers are discovering that too much screen time is linked to physical disorders, impaired social skills, and low self-esteem in the young. Children need guidance from elders, with opportunities to explore new places and ideas and to reconnect with the natural world. Children fare better when parents stay involved with their educational and recreational activities. Face-to-face interactions build social skills vital for success in life.

I urge everyone to support their local and national museums—most rely on donations and remain critically underfunded. Museums are striving to address shortened attention spans by developing new exhibits that are dynamic, interactive, and fun for inquisitive young minds. Fortify the future of tomorrow's leaders by helping museums to stay relevant in our fast-changing world.

In 1980, astronomer Carl Sagan emphasized that "You have to know the past to understand the present"—*and*, I would add, *to embrace a rewarding and sustainable future.*

A sliding window in the Prehistoric Journey exhibit at the Denver Museum of Nature and Science offers children a personalized experience with staff working in the paleontology lab. In this photo, museum volunteer Yves Genty explains how to prepare bones from a *Triceratops* skull (**above**).

Visitors to the Dominican Republic should stop at the Amber Museum in Puerto Plata (**right**), a small but beautiful display of Dominican amber housed in a colonial building.

Museums of all sizes and character offer life-enriching, educational experiences, opportunities to explore our planet's rich natural and cultural heritage. The Royal British Columbia Museum in Victoria, BC, Canada, is one of the biggies (**top right**). Its exhibits and collections span the full spectrum of natural history, with a regional focus—ranging from indigenous cultures of the Pacific Northwest to specimens from BC's world-famous fossil beds, Burgess Shale for example.

Typical theme parks are loaded with novelties for public amusement; but Mexico's Xcaret Park is a different breed. It fact, it's a world-class living natural history museum. Privately owned and operated, and located near Cozumel, Xcaret has become a tourist destination for cruise ship day-trippers. The park celebrates the ecology, cultural diversity, and archaeology of Mexico. Its exhibits are thoughtfully designed for educational, immersive experiences, with excellent interpretive signs in Spanish and English. Visitors can, among other options, swim with coral reef fish, taste traditional cuisine, explore Aztec ruins, or spend hours in a gorgeous, gargantuan aviary with waterfalls and natural habitats for over 24 species of tropical birds, mostly native to southern Mexico. It would be easy to spend a full week in this place and not see it all! Shown here is a diorama of Mayan temples (**above**).

Skeleton of a duckbilled dinosaur, *Edmontosaurus annectens*, from the Hell Creek Formation, Montana. This specimen is 80% fossil bone and 20% reconstructed. On display in the Prehistoric Journey exhibit at the Denver Museum of Nature and Science, Colorado.

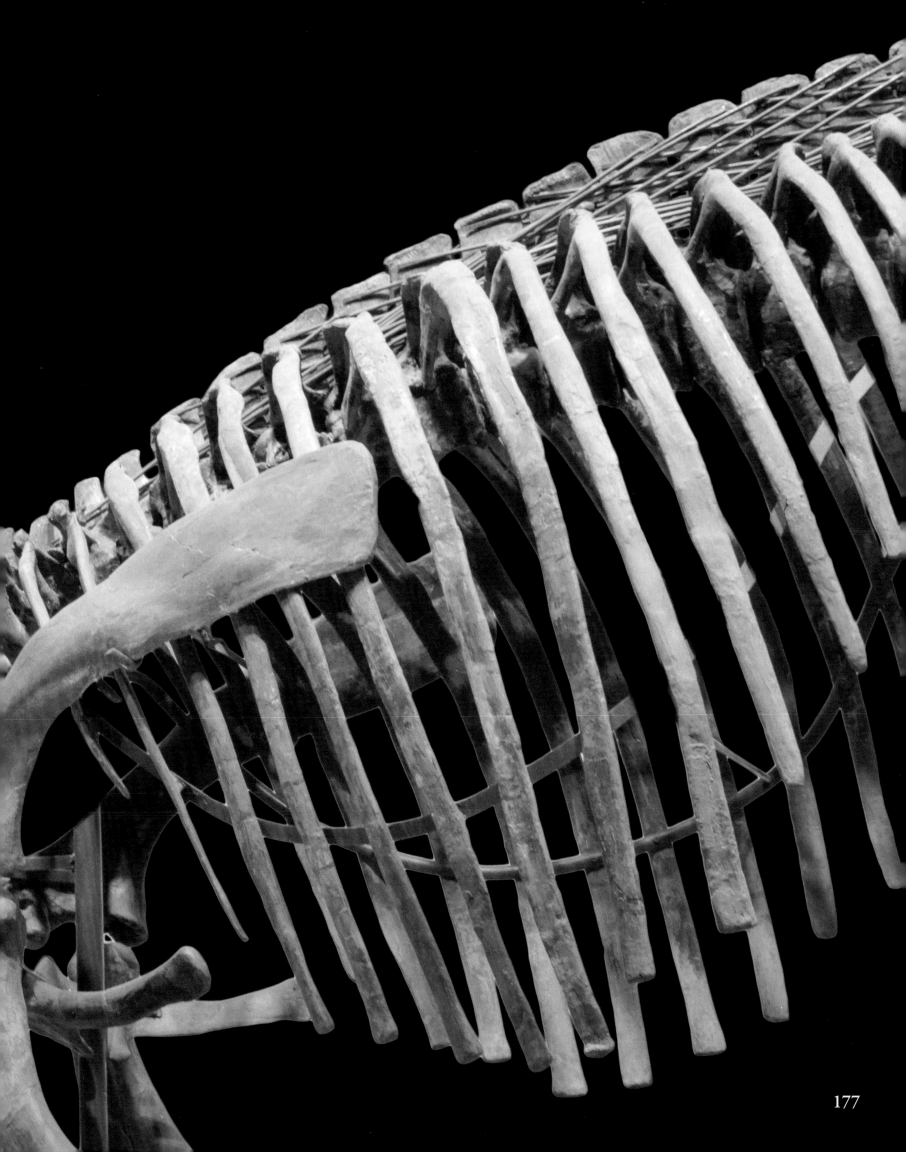

FURTHER READING

DINOSAURS & DINOSAUR HUNTERS

Bakker, Robert T. 1986. *The Dinosaur Heresies: New Theories Unlocking the Mystery of the Dinosaurs and Their Extinction*. New York: Citadel Press.

A classic by a brilliant biologist, written to convince fellow paleontologists that dinosaurs were active, intelligent, warm-blooded, bird-like creatures—a wildly unconventional view in the 1980s. Although now somewhat dated by new findings and interpretations, this educational overview is packed with Bakker's imaginative ideas and sketches.

Bakker, R. T. 1996. *Raptor Red*. Maine: Thorndike Press.

A dramatized journey into the paleo-environment and mind of an intelligent, prehistoric predator, *Utahraptor*. Bakker has used his research and revolutionary ideas about dinosaurs to create an engaging fact-with-fiction tale that offers much food for thought.

Fiffer, Steve. 2000. *Tyrannosaurus SUE*. New York: W. H. Freeman and Company.

One of the best-told stories in paleontological literature. The author has artfully woven together the discovery of T. rex SUE, the prolonged legal battle over ownership of the bones, claims and counterclaims, and vengeful prosecutions—a riveting and eye-opening tale that everyone should read.

Holtz, Thomas R. & Luis V. Rey. 2007. *Dinosaurs: The Most Complete, Up-to-Date Encyclopedia for Dinosaur Lovers of All Ages*. New York: Random House.

This beautiful guide written for young and old dino-enthusiasts—includes entries for 800+ named species of Mesozoic dinosaurs with commentaries by 33 world-famous paleontologists. But it lacks sidebars near the animals shown, so readers must hunt for descriptions in the text. A sister website is maintained with chapter updates.

Horner, John & James Gorman. 1988. *Digging Dinosaurs: The Search That Unraveled the Mystery of Baby Dinosaurs*. London: Harper Perennial.

A captivating book written for the average reader that conveys the excitement of making a paleontological find of a lifetime—hadrosaur nests, eggs, and their young at a site in Montana, the first clear evidence that dinosaurs fed and cared for their young.

Larson, Peter & Kristin Donnan. 2004. *Rex Appeal: The Amazing Story of SUE, the Dinosaur That Changed Science, the Law, and My Life*. Montpelier, VT: Invisible Cities Press.

Anyone who is even casually interested in fossils should read this book. This well-written story provides insights into hunting and collecting dinosaur remains, but more importantly, it offers personalized encounters with the FBI, Native American land owners, ego-driven prosecutors, and competitive colleagues.

Norell, Mark. 2019. *The World of Dinosaurs: The Ultimate Illustrated Reference*. Chicago, IL: Chicago Press.

This friendly and beautifully illustrated historical tour of dinosaurs on display at the AMNH—guided by museum paleontologist Mark Norell—tells readers how and where the specimens were collected, while putting them an ecological and evolutionary context. Introductory chapters cover the basics of dino-biology, followed by profiles of 44 species and the evolution of their survivors, birds.

Switek, Brian. 2015. *Prehistoric Predators*. New York: Applesauce Press.

A colorful, action-packed, large-format book that targets primary-school-age children. These dynamic illustrations by paleoartist Julius Csotonyi feature ferocious meat-eaters, mostly dinosaurs. The well-written creature profiles include a pronunciation guide and short "Science Bites."

Tennant, Jonathan, illustrated by V. Nikolov & C. Simpson. 2014. *Excavate! Dinosaurs: Paper Toy Paleontology*. North Adams, MA: Storey Publishing.

Through colorful illustrations and informative text, kids 7 and older can learn about 12 iconic dinosaurs of the Mesozoic Era. Each has a pop-out paper skeleton that can be assembled to make a stand-up model. Building them correctly requires imagination and using clues given in this book.

FOSSILS INSIDE OUT

EVOLUTION OF LIFE & FOSSIL IDENTIFICATION GUIDES

Attenborough, Sir David. 2018. *Life on Earth*. New York: HarperCollins.

This updated edition celebrates the 40th anniversary of its first publication—a companion volume to the author's ground-breaking TV series *Life on Earth*. The original text has been expanded to include modern scientific discoveries, accompanied by beautiful, all-new photographic images. This journey through time is a fine addition to any home library. Attenborough's natural history books cover many diverse topics that range from the private lives of plants to tribal art.

Eldredge, Niles. 2014. *Extinction and Evolution: What Fossils Reveal About the History of Life.* Ontario, Canada: Firefly Books.

A richly illustrated, thoughtfully constructed book with broad appeal that focuses on extinction and evolution, much of it based on research and discoveries made by the author, Curator Emeritus in paleontology at the AMNH.

Gould, Stephen Jay. 2007. *Wonderful life: The Burgess Shale and the Nature of History*. New York: Norton.

Although confusing in places for non-specialists, this book explores and explains the discovery and significance of an outstanding fossil deposit in the Canadian Rockies, the Burgess Shale, a treasure trove of bizarre Cambrian creatures.

Grande, Lance. 2013. *The Lost World of Fossil Lake: Snapshots from Deep Time*. Chicago: Univ. Chicago Press.

A clearly written, well-illustrated, authoritative, large-format reference book on fossils of Wyoming's Green River formation. About 80% of the body text is devoted to classification of fauna and flora from this locality. Back matter includes six appendices, literature citations, glossary, and index.

Johnson, Kirk R. & Richard Stucky. 2006. *Prehistoric Journey: A History of Life on Earth*. Golden, CO: Fulcrum Publishing.

Stunning photographs and artistic renderings drive this beautifully designed, updated book written by curators in earth sciences/paleoecology at the Denver Museum of Nature & Science. The informative text is inviting to the lay public.

Jones, Steve. 1999. *Almost Like a Whale: The Origin of Species Updated*. London: Doubleday.

Written by an evolutionary biologist and award-winning interpreter for the public understanding of science, the author has done a masterful job of discussing—in modern terms—Darwin's 1759 classic *On the Origin of Species*. The book is easy to read and packed with real-world examples, presented in Darwin's original chapter format. And anyone who would like to go adventuring with Darwin should get *Voyage of the Beagle*—a fascinating travelogue that's friendlier to read than his *Origin of Species*.

Larson, Neal L. 2009. *Ammonites: Treasures from a Lost World*. Tokyo, Japan: Ammolite Laboratory.

A book that will appeal to both novice and professional collectors, lavishly illustrated with color plates of specimens identified and photographed by the author. Nearly half of the book showcases ammonites from different parts of the world, and most of the rest focuses on their anatomy, evolution, and natural history.

Lebrun, Patrice. 2018. *Fossils from Morocco: Volume I. Emblematic Localities from the Paleozoic of the Anti-Atlas*. Saint-Julien-du-Pinet: Les Éditions du Piat.

An extraordinary, full-color volume, a definitive work about one of the richest fossil collecting regions in the world, with text in French and English. This 300-page tome is for paleontologists; but, text aside, it offers a stunning collection of images of fossils (many trilobites), landscapes, people, and maps.

Long, Robert, Rose Houk, & artist Doug Henderson. 1988. *Dawn of the Dinosaurs: The Triassic in Petrified Forest*. Petrified Forest, AZ: Petrified Forest Museum Association.

This popularized introduction to the prehistoric fauna and flora of Petrified Forest National Park features paintings with brief, clearly written descriptions of what is shown in each one. The beautiful artwork in this book is unusual—subdued, moody landscapes with animals, some quite small, realistically blended into the scene.

Matsen, Brad & Ray Troll. 1994. *Planet Ocean: A Story of Life, the Sea, and Dancing to the Fossil Record*. Berkeley, CA: Ten Speed Press.

More whimsical and humorous than scientifically accurate, this timeless, cartoony book is pure fun. Although not written for children, kids will enjoy the lively and colorful illustrations.

Parker, Steve. 2015. *Evolution: The Whole Story*. London: Thames and Hudson.

This hefty 576-page encyclopedic reference book could be used as a college textbook. It's a chronological compendium of prehistoric and modern life forms, with 1,000+ illustrations, biological "focal points"/sidebars, and historical commentaries. It's not a tightly focused summary of evolutionary theory that some readers might expect.

Piccini, Stefano (publisher of Nature & Discoveries series of pocket guidebooks). F. Dalla Vecchia, 2004, *The Cretaceous Fossils of Lebanon*; S. Piccini, 1997, *Fossils of the Green River Formation*; G. Poinar, 1995, *Discovering the Mysteries of Amber*. Italy: Geofin Publishing.

A beautifully produced series of handy pocket guides for fossil collectors. Each is tightly focused on one topic of interest, typically with one or two color photos per page accompanied by the specimen's identification and a brief description, with some pages devoted to geography, geology, evolution, and collecting.

Poinar, George. 1992. *Life in Amber*. Stanford, CA: Stanford Univ. Press.

This technical overview by an American entomologist begins with a detailed discussion of what amber is (vs. copal) and great amber deposits of the world. About 60% of the text is devoted to biological inclusions (especially insects), with photos, followed by chapters focused on evolution and amber research. Comprehensive references list and index.

Roberts, Alice (editor). 2018. *Evolution: The Human Story* (2nd Ed.). London: DK Publishing.

The story of human evolution is complex and fast-changing; and as of 2018, this edition had been updated with some of the latest discoveries and maps. It's primarily a picture book for adults that showcases photo-realistic illustrations by renowned Dutch paleoartists. Written by a team of experts for non-specialists.

Rose, Kenneth. 2006. *The Beginning of the Age of Mammals*. Baltimore, MD: Johns Hopkins Univ. Press.

This landmark synthesis of the early evolution and diversification of mammals is an essential reference book for paleontologists, mammologists, and graduate students in this field—the life's work of a distinguished vertebrate paleontologist. The well-written text focuses on the fossil record, supported by photographs, illustrations, and an extensive bibliography.

Schwab, Ivan R. 2012. *Evolution's Witness: How Eyes Evolved*. New York: Oxford Univ. Press.

Anyone interested is eyes, non-specialists and experts alike, should add this spectacularly illustrated, large-format book to his/her library. The author, an ophthalmologist/research biologist, has combined content from anatomy, sensory physiology, phylogeny, and the fossil record into a clear and engaging synthesis, organized by geological age.

Smithsonian (contributors). 2019. *Dinosaurs and Prehistoric Life: The Definitive Visual Guide to Prehistoric Animals*. London: DK Publishing.

This 512-page, large-format visual encyclopedia begins with the origin of planet Earth and moves through geological time, covering the full spectrum of life in the fossil record. Illustrated in color with photographs of fossils, 3D CGI imagery, charts, and simplified maps. Suitable for all age groups. Good glossary and index.

Staaf, Danna. 2020. *Monarchs of the Sea: The Extraordinary 500-Million-Year History of Cephalopods*. New York: The Experiment Publishing.

A new, paperback edition of *Squid Empire* (2017), written by a marine biologist. This engaging book is a chatty narrative well fortified with scientific insights into the evolution, anatomy, and behavior of squids, cuttlefish, octopuses, chambered nautiluses, and their extinct predecessors (especially ammonites). Black-and-white photos and illustrations.

Taylor, Paul D. & Aaron O'Dea. 2015. *A History of Life in 100 Fossils*. London: Natural History Museum.

Great in concept, this book provides an overview of prehistoric life through 100 photographs of fossils with a page of descriptive text about each, roughly arranged in geochronological order. On the downside, a few of the fossils are likely to confuse readers—some specimens are inferior and others need better explanation.

Walker, Cyril & David Ward. 2002. *Fossils: The Visual Guide to More Than 500 Species of Fossils from Around the World*. New York: DK Publishing.

A 320-page handbook is the size of a field guide, with content split evenly between invertebrate and vertebrate animals; only 25 pages are devoted to plants. Every page features photographs of two fossil specimens with helpful information panels for identification to genus, with comments on distribution. Includes a useful glossary and index.

Weiner, Jonathan. 1994. *The Beak of the Finch: A Story of Evolution in Our Time*. New York: Knopf.

A well-told introduction to the process of evolution, accessible to all readers, unburdened by tables, formulas, or jargon. Two Princeton biologists embarked on a 20-year mission in the Galápagos Islands to study evolution in living animals, mostly finches, unique to this archipelago. The timeless text is written in a conversational style, easy reading that covers many aspects of the islands' ecology, with a sprinkling of quotes from Darwin.

Witton, Mark P. 2020. *Life through the Ages II: Twenty-first Century Visions of Prehistory*. Bloomington: Indiana Univ. Press.

As a tribute to celebrated paleoartist Charles R. Knight (and his 1946 book by the same name), paleontologist/paleoartist Mark Witton begins with a discussion of how paleoscience and artistic interpretations have changed since Knight's day, followed by 62 of the author's full-color paleo-scenes with clear, concise, and scientifically credible descriptions of each. Highly recommended!

EXTINCTIONS & ENVIRONMENTAL ISSUES

Alley, Richard. 2014. *The Two-Mile Time Machine: Ice Cores, Abrupt Climate Change, and Our Future*. Princeton, NJ: Princeton Univ. Press.

Anyone concerned about global warming should read this thought-provoking book. The author, one of the world's leading climate scientists, clearly explains how climate-change data are collected, what we know and don't know about climate change, and how forecasts are made.

Alverez, Walter. 1997. *T. rex and the Crater of Doom*. Princeton, NJ: Princeton Univ. Press.

An enjoyable account of how interdisciplinary studies offered insight into Earth's mass extinction event of 66 mya, the one that wiped out about 75% of life on Earth, including dinosaurian giants.

Attenborough, Sir David, with Jonnie Hughes. 2020. *A Life on Our Planet: My Witness Statement and Vision for the Future*. New York: Grand Central Publishing.

Sir David, perhaps the greatest natural historian of our time, has spent his career traveling and documenting life in remote corners of Earth. Written at the age of 93, this biography is his "witness statement" for our planet's decline, along with his vision for the future—hopeful, but only if we act now.

Brannen, Peter. 2017. *The Ends of the World*. New York: HarperCollins.

This easy-to-read, detailed coverage of Earth's major extinction events helps readers to understand the past and how it relates to the present and future. The book is insightful and thought-provoking.

Gaul, Gilbert M. 2019. *The Geography of Risk: Epic Storms, Rising Seas, and the Cost of America's Coasts*. New York: Sarah Crichton Books.

In this carefully researched and thoughtfully written book, the author builds his case: In this era of climate change, marked by intensifying storms and coastal flooding, federal and state lawmakers are continuing to shift an unsustainable economic burden from private investors to American taxpayers.

Kolbert, Elizabeth. 2014. *The Sixth Extinction: An Unnatural History*. New York: Picador.

This book's a winner, Winner of the 2015 Pulitzer Prize. The author, a science-journalist, has filled these pages with an entertaining and thought-provoking blend of research, interviews, and fieldwork—essential reading for anyone concerned about the future of humankind.

Levitin, Daniel J. 2014. *The Organized Mind: Thinking Straight in the Age of Information Overload*. New York: Dutton.

The author, a professor of psychology and behavioral neuroscience, helps readers to understand how our brains are wired, become cluttered, and fail to perform effectively when overloaded with information—a digital age phenomenon. This book meanders in places but is full of science-based insights and self-help tips for staying focused and organized.

Louv, Richard. 2013. *Last Child in the Woods: Saving Our Children from Nature-Deficit Disorder*. London: Atlantic Books.

This important book should be read by every parent, teacher, and city planner. Citing cutting-edge research, the author clearly explains why a child's contact with nature is essential for his/her physical, emotional, and spiritual development.

MacPhee, Ross D. E. & artist Peter Schouten. 2019. *End of the Megafauna: The Fate of the World's Hugest, Fiercest, and Strangest Animals*. New York: Norton & Company.

A lucidly written, scientifically rigorous, beautifully illustrated book for scientists and non-scientists alike. It focuses on extinctions of large mammals between 5,000 and 50,000 years ago. The author carefully examines contributory causes that are rarely clear-cut and often multi-faceted.

McHarg, Ian. 2005. *Design with Nature*. New York: John Wiley & Sons.

All regional and city planners should read this award-winning classic in landscape architecture, revolutionary when first published in 1969, and still available in multiple languages. McHarg pioneered ecological planning, the virtues of which continue to be ignored by many politicians and developers.

McPhee, John. 2000. *Encounters with the Archdruid*. New York: Farrar, Straus and Giroux.

Written from a "fly-on-the-wall" perspective, the author has explored the relationship between four honorable men in three wilderness settings—each with opposing opinions on environmental issues while being forced to make tough decisions in real-life situations.

Quammen, David. 1997. *Song of the Dodo: Island Biogeography in the Age of Extinctions*. New York: Scribner.

This book is as relevant today as it was in 1997. Quammen is a gifted natural history writer and world-traveler who knows how to simplify complex concepts. He delves into history, evolution, and the origin and extinction of species in habitat islands. Everyone should read this eye-opening book!

Quammen, David. 2012. *Spillover: Animal Infections and the Next Human Pandemic.* New York: W. W. Norton.

The Covid-19 pandemic has crippled human health, social order, and economies of the world—is this "the Next Big One" discussed in Quammen's 2012 book? Written as a detective story by one of the most credible and accomplished science writers in the world, *Spillover* tracks viruses—such as the Spanish influenza of 1918, H5N1, SARS, Ebola, and HIV—that jump from wild animals to humans. Pandemics, though unpredictable, are fueled by burgeoning human and livestock populations and by what we do, or don't do.

Reisner, Marc & L. Mott. 2017. *Cadillac Desert: The American West and Its Disappearing Water*. New York, NY: Penguin Books.

In the American West, water will always be its most precious resource, and Reisner tells its story in depth—in historical, ecological, and economic terms. This thoroughly researched and engaging book is essential reading for residents and tax-payers who have settled in the West.

Sneddon, Rob. 2019. *Artificial Evolution: How Technology Makes Us Think We're Better Than We Are (and Why That's Dangerous)*. Somersworth, NH: Candlepin Press.

Anyone pondering the future of humanity should read this book. In engaging, non-technical terms, the author considers the consequences of modern technology, its pace, and delusions that come with it, framed in the context of evolution.

Steinmetz, George & Andrew Revkin. 2020. *The Human Planet: Earth at the Dawn of the Anthropocene*. New York: Abrams Press.

A magnificent large-format collection of full-page aerial photographs from around the globe. These graphic images and their informative captions help us to envision the future of humanity, as we struggle to provide shelter, food, water, and energy for our burgeoning populations.

Wood, Gillen D'Arcy. 2015. *Tambora: The Eruption that Changed the World*. Princeton, NJ: Princeton Univ. Press.

Although guilty of repetition and wandering off-topic in places, this account of the worst volcanic explosion in recorded history—with its long-lasting global impact on human health, agriculture, and economic well-being—gives readers a lot to think about in this era of global warming.

FLORA & FUNGI

Armstrong, Joseph E. 2015. ***How the Earth Turned Green: A Brief 3.8-Billion-Year History of Plants.*** Chicago: Univ. Chicago Press.

This book is full of useful and surprising information, written for general readers. The author explores the diversity of simple life forms, pioneering land plants, early forests, the rise of flowering plants, and the importance of plants to humans.

Ash, Sydney. 2005. ***Petrified Forest: A Story in Stone***. Petrified Forest, AZ: Petrified Forest Museum Association.

A beautifully conceived and designed overview of the past and present of Petrified Forest National Park, richly illustrated with drawings and color photographs. This short, large format book is easy reading, but has no index.

Buchmann, Stephen. 2015. ***The Reason for Flowers: Their History, Culture, Biology, and How They Change Our Lives***. New York: Scribner.

An easy-read, a broad mix of biological and cultural information about flowering plants, with snippets about everything from plant breeding and floral teas to perfumes and funerals, supported by a 19-page index. Pollination ecology is a recurring theme, especially bees—the author's specialty. No color illustrations.

Kassinger, Ruth. 2019. ***Slime: How Algae Created Us, Plague Us, and Just Might Save Us***. Boston: Houghton Mifflin Harcourt.

A mind-bending, fun-to-read book that focuses on how an under-appreciated group of organisms have created the world we know today and might attain tomorrow. These often-ignored "lowly" photosynthesizers are full of surprises.

Kress, W. John & Shirley Sherwood. 2009. ***The Art of Plant Evolution. Royal Botanic Gardens***. Kew, England: Kew Publishing.

This beautiful blend of art and science begins with a 22-page introduction to botanical art, natural selection, co-evolution, and classification based on recent DNA data. Most of the book is devoted to 136 full-page paintings, each with a brief botanical description, comments about the artist, and notes about the group of plants to which it belongs.

Sheldrake, Merlin. 2020. ***Entangled Life: How Fungi Make Our Worlds, Change Our Minds & Shape Our Futures***. New York: Random House.

Enter the mystical world of fungi through this fascinating account of their biology, ecology, and psycho-physiology. Unfortunately, Sheldrake ignores the dark side of hallucinogens, and his book suffers from an annoying amount of repetition.

GEOLOGY & PLANETARY SCIENCE

Bjornerud, Marcia. 2008. ***Reading the Rocks: The Autobiography of the Earth.*** Paradise, CA: Paw Prints.

The author is a gifted geo-scientist and story-teller, who explains the rich history of Earth through deep time as revealed in rocks and fossils. This somewhat poetic book is more about principles and processes than facts and numbers. Short on illustrations but contains a helpful glossary and timescale.

Norton, Richard O. 2004. ***Rocks from Space: Meteorites and Meteorite Hunters.*** Missoula, MT: Mountain Press Publishing.

Above all, this is a comprehensive non-technical, well-illustrated book about meteorites, with historical notes—written to help meteorite enthusiasts find, identify, understand, and classify them. Topics covered in less detail are impact craters, new discoveries in astronomy, asteroids, comets, and extinction events. Excellent glossary, references, and index.

Schulze-Makuch, Dirk and William Bains. 2017. ***The Cosmic Zoo: Complex Life on Many Worlds***. Cham, Switzerland: Springer Nature.

The authors, both university professors and astrobiologists, explore an age-old question—how likely is the evolution of complex life elsewhere in the Universe? Their approach is based on what we know about the biology and evolution of life on Earth. This 200-page book is thought-provoking and easily digested by readers with no scientific training.

Vogel, Shawna. 2001. ***Naked Earth: The New Geophysics***. New York: Plume.

The author, a skilled science reporter for *Discover Magazine*, clearly explains the dynamic character of our planet, starting with its structure and the discovery of plate tectonics. She explores geophysical research, super-continents, volcanoes, magma plumes, the atmosphere, and Earth as part of our solar system.

HISTORICAL PERSPECTIVES & BIOGRAPHIES

Cadbury, Deborah. 2000. *The Dinosaur Hunters: A True Story of Scientific Rivalry and the Discovery of the Prehistoric World*. London: HarperCollins.

A well-told story of bitter rivalry between two prominent British 19th-century scientists, eccentric men who made amazing discoveries, tainted by unrelenting greed.

Carrano, M. T., K. R. Johnson, & illustrator Jay Matternes. 2019. *Visions of Lost Worlds: The Paleoart of Jay Matternes*. Washington, D.C.: Smithsonian Books.

A beautiful book honoring the career of an exceptionally talented paleoartist, Jay Matternes, who has painted murals and dioramas for major museums in the USA. This book illustrates the process of creating seven amazing murals in detail, including his anatomical study sketches and natural history notes about each scene.

Emling, Shelley. 2009. *The Fossil Hunter: Dinosaurs, Evolution, and the Woman Whose Discoveries Changed the World*. New York: Palgrave Macmillan.

A biography of self-taught paleontologist Mary Anning (1799-1847), whose passion for science and fossil-hunting along the Lyme Regis cliffs of England was hindered by poverty and a struggle for recognition—at that time, only men of means were considered worthy of a career in science. A 2020 British film, *Ammonite*, tells her story, spiced-up with a lesbian romance having no known historical basis.

Jaffe, Mark. 2000. *The Gilded Dinosaur: The Fossil War between E. D. Cope and O. C. Marsh and the Rise of American Science*. New York: Crown Publishers.

Set in the second half of the 19th century, this engaging historical account of what began as a friendly quest for dinosaur bones by two brilliant young American paleontologists evolved into a vicious ego-driven rivalry that lasted for decades.

Weinberg, Samantha. 2001. *A Fish Caught in Time: The Search for the Coelacanth*. New York: Harper Perennial.

This well-told detective story is light reading, an intriguing historical account of finding live specimens of an ancient lineage of fish with limb-like fins, known only from fossils before 1938. Its significance fueled rivalries among scientists and politicians in pursuit of fame.

Willmann, Rainer & Julia Voss. 2020. *The Art and Science of Ernst Haeckel.* Köln, Germany: Taschen GmbH (translated from a 2017 German edition).

Ernst Haeckel needs no reintroduction here (see pp. 4-5, 92-93, 134-135, 142-143, 157, & 185 in *Fossils Inside Out*). This 512-page tribute to his life and work is small—6.5 x 8.5 inches (16 x 21.5 cm)—but a beautifully printed volume with more than 300 of Haeckel's finest prints, well supported by text written by two experts. Rainer is co-founder of Göttingen University's Center for Biodiversity & Ecology Research and Voss' specialty is visual representation of Darwinian evolution. Note: a much larger, expensive, 2019 edition is available (ISBN 9783836526463).

Wilson, E. O. 2006. *Naturalist*. Washington, D.C.: Island Press.

An inspiring autobiography by one of the world's pre-eminent ecologists and evolutionary biologists, a two-time Pulitzer Prize winner. Starting with his childhood in Alabama, Wilson traces his intellectual development and challenges faced during his 40-year tenure as a research professor at Harvard—best known for his work on social insects, biodiversity, and sociobiology. Every budding scientist should read this book.

Winchester, Simon & photographer Soun Vannithone. 2016. *The Map That Changed the World: William Smith and the Birth of Modern Geology*. New York: MJF Books.

A biography of British engineer-geologist William Smith (1769-1839), the first to map fossils in stratigraphic order. Here's an inspiring story of a man rejected by the better-educated scientific establishment for most of his career, and later recognized as the "Father of English Geology."

Wulf, Andrea. 2015. *The Invention of Nature: Alexander von Humboldt's New World*. New York: Knopf.

A captivating, meticulously researched biography of a famous 19th-century New World explorer/naturalist/visionary. Humboldt's books influenced many scientists, poets, and politicians—Darwin, Haeckel, Thoreau, Emerson, Muir, & Jefferson among them. He denounced colonialism and slavery. A full 27% of this book is devoted to endnotes and an index.

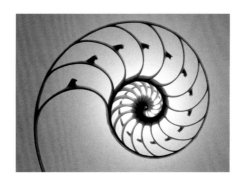

Nothing is constant but change!
All existence is a perpetual flux of "being and
becoming!" That is the broad lesson of the
evolution of the world, taken as a whole
or in its various parts.

~ from *THE WONDERS OF LIFE,*
A Popular Study of Biological Philosophy
by Ernst Haeckel (1905)—
a 2014 Project Gutenberg translation